IET CONTROL ENGINEERING SERIES 90

Application of Dimensional Analysis in Systems Modeling and Control Design

Other volumes in this series:

Volume 8	**A History of Control Engineering, 1800–1930** S. Bennett
Volume 18	**Applied Control Theory, 2nd Edition** J.R. Leigh
Volume 20	**Design of Modern Control Systems** D.J. Bell, P.A. Cook and N. Munro (Editors)
Volume 28	**Robots and Automated Manufacture** J. Billingsley (Editor)
Volume 33	**Temperature Measurement and Control** J.R. Leigh
Volume 34	**Singular Perturbation Methodology in Control Systems** D.S. Naidu
Volume 35	**Implementation of Self-tuning Controllers** K. Warwick (Editor)
Volume 37	**Industrial Digital Control Systems, 2nd Edition** K. Warwick and D. Rees (Editors)
Volume 39	**Continuous Time Controller Design** R. Balasubramanian
Volume 40	**Deterministic Control of Uncertain Systems** A.S.I. Zinober (Editor)
Volume 41	**Computer Control of Real-time Processes** S. Bennett and G.S. Virk (Editors)
Volume 42	**Digital Signal Processing: principles, devices and applications** N.B. Jones and J.D.McK. Watson (Editors)
Volume 44	**Knowledge-based Systems for Industrial Control** J. McGhee, M.J. Grimble and A. Mowforth (Editors)
Volume 47	**A History of Control Engineering, 1930–1956** S. Bennett
Volume 49	**Polynomial Methods in Optimal Control and Filtering** K.J. Hunt (Editor)
Volume 50	**Programming Industrial Control Systems Using IEC 1131-3** R.W. Lewis
Volume 51	**Advanced Robotics and Intelligent Machines** J.O. Gray and D.G. Caldwell (Editors)
Volume 52	**Adaptive Prediction and Predictive Control** P.P. Kanjilal
Volume 53	**Neural Network Applications in Control** G.W. Irwin, K. Warwick and K.J. Hunt (Editors)
Volume 54	**Control Engineering Solutions: a practical approach** P. Albertos, R. Strietzel and N. Mort (Editors)
Volume 55	**Genetic Algorithms in Engineering Systems** A.M.S. Zalzala and P.J. Fleming (Editors)
Volume 56	**Symbolic Methods in Control System Analysis and Design** N. Munro (Editor)
Volume 57	**Flight Control Systems** R.W. Pratt (Editor)
Volume 58	**Power-plant Control and Instrumentation: the control of boilers and HRSG systems** D. Lindsley
Volume 59	**Modelling Control Systems Using IEC 61499** R. Lewis
Volume 60	**People in Control: human factors in control room design** J. Noyes and M. Bransby (Editors)
Volume 61	**Nonlinear Predictive Control: theory and practice** B. Kouvaritakis and M. Cannon (Editors)
Volume 62	**Active Sound and Vibration Control** M.O. Tokhi and S.M. Veres
Volume 63	**Stepping Motors, 4th Edition** P.P. Acarnley
Volume 64	**Control Theory, 2nd Edition** J.R. Leigh
Volume 65	**Modelling and Parameter Estimation of Dynamic Systems** J.R. Raol, G. Girija and J. Singh
Volume 66	**Variable Structure Systems: from principles to implementation** A. Sabanovic, L. Fridman and S. Spurgeon (Editors)
Volume 67	**Motion Vision: design of compact motion sensing solution for autonomous systems** J. Kolodko and L. Vlacic
Volume 68	**Flexible Robot Manipulators: modelling, simulation and control** M.O. Tokhi and A.K.M. Azad (Editors)
Volume 69	**Advances in Unmanned Marine Vehicles** G. Roberts and R. Sutton (Editors)
Volume 70	**Intelligent Control Systems Using Computational Intelligence Techniques** A. Ruano (Editor)
Volume 71	**Advances in Cognitive Systems** S. Nefti and J. Gray (Editors)
Volume 72	**Control Theory: a guided tour, 3rd Edition** J.R. Leigh
Volume 73	**Adaptive Sampling with Mobile WSN** K. Sreenath, M.F. Mysorewala, D.O. Popa and F.L. Lewis
Volume 74	**Eigenstructure Control Algorithms: applications to aircraft/rotorcraft handling qualities design** S. Srinathkumar
Volume 75	**Advanced Control for Constrained Processes and Systems** F. Garelli, R.J. Mantz and H. De Battista
Volume 76	**Developments in Control Theory towards Glocal Control** L. Qiu, J. Chen, T. Iwasaki and H. Fujioka (Editors)
Volume 77	**Further Advances in Unmanned Marine Vehicles** G.N. Roberts and R. Sutton (Editors)
Volume 78	**Frequency-Domain Control Design for High-Performance Systems** J. O'Brien
Volume 81	**Optimal Adaptive Control and Differential Games by Reinforcement Learning Principles** D. Vrabie, K. Vamvoudakis and F. Lewis
Volume 88	**Distributed Control and Filtering for Industrial Systems** M. Mahmoud
Volume 91	**An Introduction to Fractional Control** D. Valério and J. Costa

Application of Dimensional Analysis in Systems Modeling and Control Design

Pedro Balaguer

The Institution of Engineering and Technology

Published by The Institution of Engineering and Technology, London, United Kingdom

The Institution of Engineering and Technology is registered as a Charity in England & Wales (no. 211014) and Scotland (no. SC038698).

First published 2013

The Institution of Engineering and Technology
Michael Faraday House
Six Hills Way, Stevenage
Herts, SG1 2AY, United Kingdom

www.theiet.org

British Library Cataloguing in Publication Data
A catalogue record for this product is available from the British Library

ISBN 978-1-84919-621-5 (hardback)
ISBN 978-1-84919-622-2 (PDF)

Typeset in India by MPS Limited
Printed in the UK by CPI Group (UK) Ltd, Croydon

Contents

Preface ix

1 Introduction **1**
1.1 What is dimensional analysis? 1
1.2 What is dimensional similarity? 2
1.3 Application of dimensional analysis to science in general 3
1.3.1 Structure of physical relations 3
1.3.2 Dimensionless representation 5
1.3.3 Dimensional similarity 5
1.4 Application of dimensional analysis to control problems 6
1.4.1 Identification and model validation 8
1.4.2 Control theory 8
1.4.3 Control engineering 8
1.5 Book contents 9

2 Dimensional analysis and dimensional similarity **11**
2.1 Physical quantities, units, and dimensions 11
2.1.1 Physical quantity 11
2.1.2 Units 11
2.1.3 Dimensions: fundamental and derived 11
2.1.4 Arithmetic of dimensions 12
2.2 Systems of units: dependence and independence of dimensions 13
2.2.1 System of units 13
2.2.2 Monomial power law 14
2.2.3 Dependent and independent dimensions 14
2.3 Buckingham pi theorem 14
2.4 Matrix approach for finding the dimensionless numbers 16
2.4.1 The dimensional matrix 16
2.4.2 The dimensional set 17
2.5 Dimensional similarity 18
2.5.1 Scale factors 20
2.5.2 Model law 21
2.6 Exercises 21
References 22

3 Dynamical systems: dimensionless representation **23**
3.1 Introduction 23
3.2 Transfer function dimensionless representation 23
3.2.1 Transfer function parameter dimensions 24
3.2.2 Transfer function parameters with independent dimensions 24
3.2.3 Transfer function dimensionless numbers 25
3.2.4 Dimensionless transfer function 26
3.3 State space dimensionless representation 27
3.3.1 Interpretation of the state space dimensionless transformation 30
3.4 Comparison between transfer function and state space dimensionless representation 32
3.5 Discrete time models dimensionless representation 35
3.5.1 Discrete time transfer function dimensionless representation 35
3.5.2 Discrete time state space model dimensionless representation 36
3.6 Exercises 38
References 38

4 Dynamical systems: dimensional similarity **41**
4.1 Introduction 41
4.2 Continuous time dynamical systems similarity 41
4.2.1 Transfer function dimensional similarity 41
4.2.2 State space dimensional similarity 44
4.3 Discrete time dynamical system similarity 46
4.3.1 Discrete time transfer function similarity 46
4.3.2 Sampled-data transfer function similarity 46
4.3.3 Discrete state space similarity 47
4.4 Exercises 48
References 49

5 Dimensionless systems identification and model order reduction **51**
5.1 Introduction 51
5.2 General procedure 51
5.3 Example 1: Second order inverse response model identification 52
5.3.1 Problem statement 52
5.3.2 Dimensionless representation of second order inverse response model 53
5.3.3 Identification procedure 54
5.3.4 Application examples 61
5.4 Example 2: Reduced effective transfer function reduction for PID decentralized control 64

5.4.1 Problem statement 64
5.4.2 Dimensionless representation of the reduced effective transfer function 67
5.4.3 Inverse response analysis 68
5.4.4 Reduced order model: general case 72
5.4.5 Reduced order model: particular cases 73
5.4.6 Application examples 75
References 77

6 Homogeneity of PID tuning rules 79
6.1 Introduction 79
6.2 Homogeneous PID tuning rules 80
6.2.1 Dimensionless controller parameters 80
6.2.2 Homogeneous tuning rules characterization 80
6.2.3 Dimensionless controller representation with homogeneous tuning rules 83
6.3 Closed loop transfer functions 83
6.3.1 Loop transfer function *GC* 83
6.3.2 Dimensionless closed loop transfer functions 85
6.4 Optimality of homogeneous tuning rules 87
6.4.1 Weighting factors 88
6.5 Homogeneous and nonhomogeneous tuning rules 89
References 94

7 Dimensionless PID tuning rules comparison 97
7.1 Introduction 97
7.2 Elements of the comparative framework 98
7.3 Dimensionless comparative framework 99
7.4 Dimensionless elements 100
7.4.1 Loop transfer function *GC* 100
7.4.2 Dimensionless closed loop transfer functions 100
7.4.3 Dimensionless integral errors 100
7.4.4 Indexes 102
7.5 Application example 102
7.5.1 PID tuning rules dimensionless characterization 103
7.5.2 Dimensionless sensitivity bandwidth comparison w_b 103
7.5.3 Dimensionless sensitivity peak comparison 104
7.5.4 Dimensionless integral absolute error 104
7.5.5 Dimensionless control action variation 106
7.6 PID tuning rules selection 106
References 108

8 Control of dimensionally similar systems 109
8.1 Introduction 109

8.2 Control of dimensionally similar systems 110
8.3 Complete similarity 111
8.3.1 Continuous time control 111
8.3.2 Discrete time control 111
8.4 Partial similarity 113
8.5 Experimental case study 115
References 119

9 Adaptive systems 121
9.1 Introduction 121
9.2 Actuator limitations and dimensionally similar model reference 122
9.2.1 Control effort 122
9.2.2 Similar model reference adaptive control 123
9.3 SMRAC for first order plants 126
9.4 SMRAC for arbitrary order plants 129
9.4.1 SMRAC control scheme 129
9.4.2 SMRAC stability analysis 131
9.4.3 SMRAC operation modes 133
9.5 Application example 134
References 137

Preface

Purpose of the book

This book is about applying dimensional analysis concepts to control theory and control engineering problems. We find two fundamental book objectives: (i) to introduce the fundamentals of dimensional analysis to control engineers and researchers and (ii) to show how to exploit the benefits that dimensional analysis can provide to control theoretic and practical problems.

The rationale for writing the book lies in the novelty of the topic for the control community and its applicability to control engineering problems. As a result, this book is intended to familiarize the reader with the basic concepts, principles, and methods of dimensional analysis and its applications to modeling and control design. Since the book is introductory in nature, we have considered control theory applications at senior undergraduate level or beginning graduate students. In particular we have focused on modeling and PID process control. More specific chapters on particular fields on control systems design, such as robust control of dimensionally similar systems and adaptive control, require previous knowledge on the fundamentals of those fields.

Prerequisites

The book is elementary. A background in undergraduate mathematics and physics, in particular linear algebra, ordinary calculus, and physic fundamentals, is sufficient as a prerequisite for the first half of the book. Applications of modeling and PID process control may also be followed by undergraduate students with a process control related minor. The book should therefore be accessible to a wide spectrum of students. Control engineers may also follow the great majority of chapters related to process control.

Content and arrangement

The content of the book has been arranged in four major blocks. *Dimensional analysis fundamentals*, covered by Chapters 1 and 2, motivates the use of dimensional analysis and introduces the basic concepts of dimensional analysis. Next, we develop the *Application of dimensional analysis and dimensional similarity to systems and control*, in Chapters 3 and 4, where fundamental results are obtained for the dynamical system dimensionless representation and the characterization of dimensionally similar systems. *Dimensional analysis and process control* is covered in Chapters 5–7. Dimensional analysis is applied to system identification and model order reduction, the introduction of the homogeneity concept of PID tuning rules, and the development of a framework for homogeneous PID tuning rules comparison. Finally, *New applications of dimensional analysis to advanced control systems*, covered by Chapters 8 and 9, presents the potential research applications of dimensional analysis to advanced control systems such as robust control of dimensionally similar systems and adaptive control.

Acknowledgments
This book has been the result of the research work of many people. First of all, we are in debt to Dr. Andrew G. Alleyne, Dr. Sean Brennan, and Dr. Haftay Haily for their initiative and development in the application of the dimensional analysis to the control field.

In particular, I would like to thank Víctor Alfaro for his hospitality, and Orlando Arrieta and Julio Romero for their friendship. Chapter 5 would not have been possible without them. I am also in debt to my former colleagues Asier Ibeas, Carles Pedret, and Salvador Alcántara. Their knowledge and wisdom have made possible Chapter 6. Finally I would like to thank Alejandro Claramonte for his intense work and dedication in some parts of Chapter 8.

The author wishes to thank Dr. Nigel Hollingworth for his interest, help, and support, making possible this work, and to Ting Ting Liu for her help and patience with the book development. I would like to thank also Dr. Reza Katebi for his guidance, wise counseling, and friendship.

Finally, I would like to thank my wife Maria for her patience while writing this book. I dedicate this book to her and to my daughter Mireia.

Chapter 1
Introduction

1.1 What is dimensional analysis?

Scientific theories are inventions of the human mind that explain and predict phenomena. The development of scientific theories uses mathematics as a fundamental tool to provide a formal framework for deductive reasoning. Scientific theories that explain physical relations deal with physical dimensions. In this case, the mathematical framework is enriched with another axiom, derived from a very simple but fundamental idea: physical laws do not depend on the arbitrariness in the choice of basic units of measurements, that is physical laws are *homogeneous*.

A mathematical equation is valid if both sides of the equality sign are equal. A physical equation also requires that both sides of the equality sign have the same physical dimensions, that is the equation must be homogeneous. As a result, a physical equation, for validity, requires not only mathematical equality but also dimensional homogeneity.

In the realm of mathematics the equation

$$x^2 = x \tag{1.1}$$

is perfectly valid and has the solutions $x = 0$ and $x = 1$. However, consider now that equation (1.1) has a physical meaning. For instance variable x is the distance between two points, thus it has length dimension, that is $[x] = L$. In this case, equation (1.1) is not a dimensionally homogeneous equation because the left hand side has area dimension, $[x^2] = L^2$, whereas the right hand side has length dimension, $[x] = L$. We may conclude that physical equations that are not dimensionally homogeneous are not physically valid because in case of non-homogeneity the physical magnitudes would be dependent on the choice of the measurements units.

The origins of dimensional analysis may be tracked back to several centuries, although its foundations were established two centuries ago and it was the last century that witnessed the application of dimensional analysis to a wide variety of physical problems. One of the difficulties found in developing the dimensional analysis was the existence of derived quantities, which were not established until the works of Newton and Leibniz around the beginning of the eighteenth century.

At the end of the nineteenth century the advances of dimensional analysis were applied to fluid dynamic problems, which had a difficult analytic solution, such as the works of Helmholtz on hydrodynamics and Rayleigh on the theory of sound.

Finally, the fundamental theorem of dimensional analysis was established independently by Federmann, Riabouchinsky, and Buckingham at the beginning of the twentieth century. This fundamental theorem, known as the Buckingham pi theorem, states that a physical relation can be reformulated in a dimensionless form, in such a way that the number of dimensionless parameters is less than or equal to the number of dimensional parameters. The major implication of the theorem is that for any physical relation there exists a compressed representation because it has the fewest required parameters. Furthermore, these parameters are dimensionless.

Summing up, dimensional analysis is based on the homogeneity property of physical equations. Homogeneity is a further constraint on mathematical equations that is a necessary condition for correctness of physical equations. The fundamental result of dimensional analysis is the Buckingham pi theorem, which establishes the existence of dimensionless representation of physical equations. The dimensionless representation is useful in several ways, one of them is to establish dimensional similarity between systems.

1.2 What is dimensional similarity?

The use of scaled models is prevalent in engineering applications because they allow performance of experimental tests without requiring the physical system (i.e. prototype) to be constructed. The use of models is beneficial because it allows experimental data to be obtained that would be very difficult to obtain analytically or computationally, and, at the same time, provides safety and economic advantages.

However, to extrapolate model experimental data to prototype in a meaningful way, the model must be built dimensionally similar with respect to the prototype. One trait of dimensional similarity is geometrical similarity, as can be seen in Figure 1.1. However, there are other forms of dimensional similarity such as kinematic similarity or thermal similarity.

The existence of dimensional similarity is also related to the existence of scaling factors between model and prototype. In fact it is the existence of the scaling factors that permits the data extrapolation from model to prototype. A scaling factor with respect to a particular physical variable is the quotient of the magnitudes of that variable for the prototype and for the model. It is a common misconception to talk about the model scaling factor as if it was unique. From the scaling factor definition

Figure 1.1 Model and prototype showing geometrical similarity

it follows that there are as many scaling factors as physical parameters involved in the model experimentation.

The fundamental Buckingham pi theorem of dimensional analysis establishes that any physical relation can be written in dimensionless form by using dimensionless parameters. The dimensionless representation of physical systems establishes a formal baseline for dimensional similarity characterization: two systems, model and prototype, are dimensionally similar if both systems have the same dimensionless representation. In this way the dimensionless numbers are the fingerprints of dimensional similarity.

The use of models has also a long history. It began in the sixteenth century with the use of models for investigating the strength of materials for buildings and the construction of machines. At the end of the nineteenth century dimensional similarity was systematically applied to solve hydrodynamic problems, giving rise to several famous dimensionless numbers such as the Froude number, the Reynolds number, and the Prandtl number, among others. In fact, the fluid dynamic science uses models extensively because the problems faced have a difficult analytical solution and are computationally very demanding. However, the use of models is pervasive in any area of engineering such as mechanics, thermodynamics, and chemistry.

1.3 Application of dimensional analysis to science in general

Once the concepts of homogeneity, dimensional analysis, and dimensional similarity are established, we introduce the applications of dimensional analysis to science in general. These are:

1. *Structure of physical relations:* The homogeneity property may be helpful to find the structure of physical relations.
2. *Dimensionless representation:* The Buckingham pi theorem provides a compact representation of physical relations through dimensionless numbers. This is beneficial because:
   - It helps to understand the phenomenon under study, not only the depending quantities but also their relations.
   - The results can be presented in a more compact way with fewer variables.
3. *Dimensional similarity:* If we can establish dimensional similarity among several objects of study then:
   - The experimental results can be generalized.
   - The number of experiments can be minimized.

1.3.1 Structure of physical relations

Physical relations are derived through experimentation and/or analytical derivation from first principles. However, in some cases, the physical relation can be derived without any prior knowledge or experimentation, by only applying the homogeneity property to the physical parameters involved in the phenomenon under study.

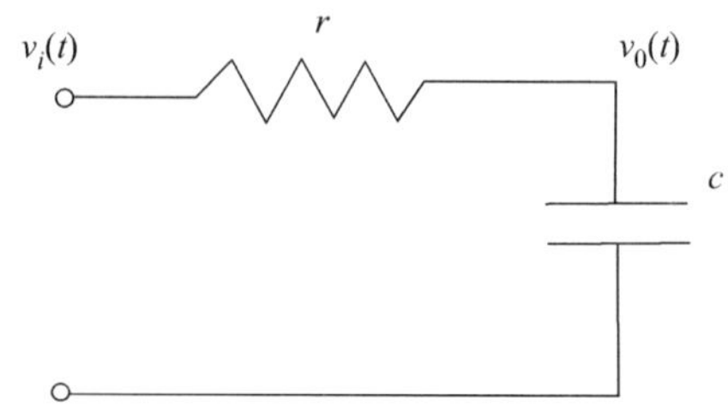

Figure 1.2 Circuit showing the capacitor charge through a resistance. $v_i(t)$ is the voltage input, v_0 is the capacitor charge, c is the capacitor capacitance, and r is the electrical resistance

A classical example is the derivation of the period t_p of a single pendulum of length l, in the presence of gravitational acceleration g. By looking for a homogeneous equation it is possible to find that $t_p = k\sqrt{l/g}$, with k an unknown constant.

We show the procedure of finding the structure of a physical relation by means of an electrical problem. Consider the simple circuit shown in Figure 1.2, which shows the charge of a capacitor through a resistance. Let's consider that the time needed to charge capacitor t_c is a function of the capacitor capacitance c and the resistance r. Mathematically we express $t_c = f(r, c)$ and we want to find the form of the function $f(\cdot, \cdot)$. The units of the involved parameters are $[t_c] = T$, $[c] = AT/V$, and $[r] = V/A$, where T is time, V is voltage, and A is electrical current.

Recalling that the product rc has time dimension, that is $[rc] = T$, it is easy to derive the following dimensionless number using the three involved parameters:

$$\Pi = \frac{t_c}{RC} \tag{1.2}$$

Π must be constant under any arbitrary dimensions used to measure time, voltage, and current, due to its dimensionless nature. As a result, (1.2) can be written as

$$t_c = \Pi rc \tag{1.3}$$

with Π a dimensionless constant. It follows that the time needed to charge a capacitor is proportional to the capacitance multiplied by the resistance. Therefore, we have established that (i) the function $f(\cdot, \cdot)$ does not depend on two parameters but just on one parameter rc, and (ii) the dependence is linear through an unknown constant Π. It is important to stress that we have arrived at the conclusion by just applying the homogeneity principle and no physical knowledge has been applied beyond the units of the parameters involved.

In the preceding discussion we have not considered the voltage input v_i nor the initial capacitor charge v_0. Moreover, the capacitor charge time t_c does not explicitly define the final charge v_f to be reached. However, despite these not considered factors, if they are held constant, it is true that the capacitor charge time is given by $t_c = \Pi rc$, and the value of the dimensionless constant Π is given by $\Pi = g(v_i, v_0, v_f)$, with g an unknown function.

1.3.2 Dimensionless representation

The existence of a dimensionless representation allows for the understanding of the phenomenon under study beyond the isolated enumeration of physical parameters involved. Following the previous electrical example shown in Figure 1.2, we ask on the evolution of capacitor voltage v_0, when the voltage input is a step of value v_i. We envisage that mathematically $v_0(t) = f(t, v_i, r, c)$, that is the voltage output v_0 is a function of time t, the amplitude of step v_i, the capacitance c, and the resistance r. Of course $v_0(t)$ can be analytically obtained by solving a differential equation, but we want to obtain some insight by mere dimensional evaluation.

First we form the following dimensionless numbers: $v_0(t)/v_i$ and t/rc. In this way the dimensional expression $v_0(t) = f(t, v_i, r, c)$ can be expressed in dimensionless form as

$$\frac{v_0(t)}{v_i} = \phi\left(\frac{t}{rc}\right) \tag{1.4}$$

with ϕ an unknown function. It follows that $v_0(t)$ depends on just one parameter, the dimensionless time t/rc. However, the dependence of $v_0(t)$ with respect to v_i is just linear as easily seen by rewriting previous expression as

$$v_0(t) = v_i\phi\left(\frac{t}{rc}\right) \tag{1.5}$$

The amount of information provided in (1.5) by mere dimensionless representation is of great significance because the parameters' dependence on $v_0(t)$ has been reduced from 4 to 2, and one of them shows a linear relationship.

Another application of the dimensionless representation is to present experimental results in a more compact way. Consider for instance that a series of experiments is performed to investigate the form of the function $v_0(t) = f(t, v_i, r, c)$. In this way eight combinatorial experiments are performed with the following values: $r = \{1, 2\}$ Ohms, $c = \{1, 3\}$ Farads, and $v_i = \{1, 2\}$ Volts. The experimental results can be seen in Figure 1.3(a).

Now, if we take the same eight experiments but represent them in dimensionless form by (i) scaling the voltage output $v_0(t)/v_i$ and (ii) scaling the time axis by t/rc, we obtain Figure 1.3(b). It must be stressed that Figure 1.3(b) is the representation of the eight experiments, which all collapse in a single graph, showing the same experimental results in just one graph.

1.3.3 Dimensional similarity

If dimensional similarity can be established between physical systems, the experimental results can be generalized. For instance the capacitor charge $v_0(t)$ follows the dimensional equation

$$v_0(t) = (1 - e^{-t/rc})v_i \tag{1.6}$$

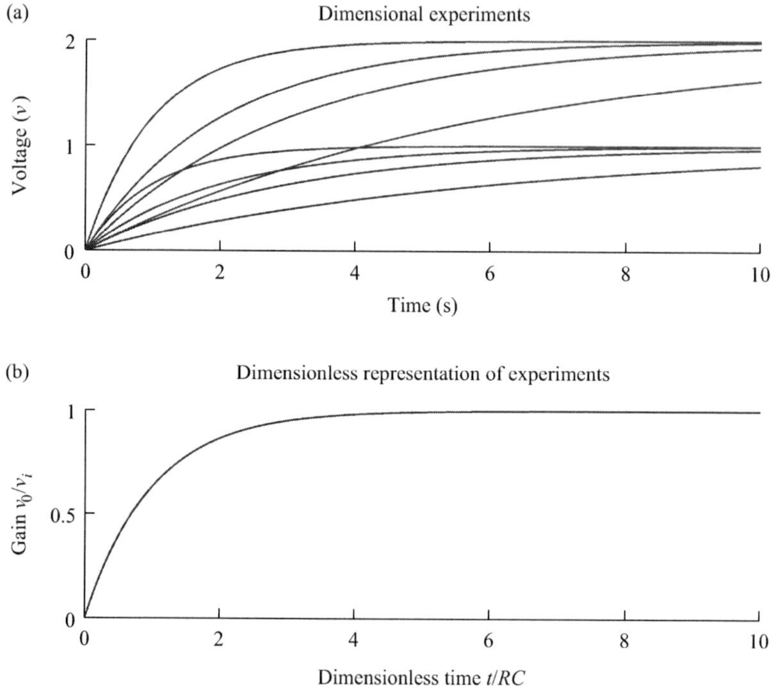

Figure 1.3 Dimensional and dimensionless representations of experiments. (a) The eight experimental results for the capacitor charge, for r = {1, 2} Ohms, c = {1, 3} Farads, and v_i = {1, 2} Volts. (b) The same experimental results as a function of dimensionless variables. In dimensionless form the eight experiments provide the same result

which can be written in dimensionless form as

$$v_0(t)/v_i = 1 - e^{-t/rc} \tag{1.7}$$

Dimensionless equation (1.7) depends only on one dimensionless number t/rc. Recalling that two systems are dimensionally similar when they have the same dimensionless numbers, it can be deduced that for any values of r and c, a time scale that makes the systems dimensionally similar can be calculated. Consequently, the experimental results can be generalized. In this way, the number of experiments can be minimized. In fact, in the previous capacitor voltage experiment, only one experiment is required and the capacitor voltage evolution can be easily obtained for any value of v_i, r, and c by amplitude and time scaling, as previously shown.

1.4 Application of dimensional analysis to control problems

The objective of control engineering is to make a physical system to behave in a specified way, defined normally by some performance measure. The types of systems to be controlled are enormous and each day new applications to control are envisaged, designed, and applied. For example control applications range from classical fields such as mechanical systems, electronic devices, and chemical processes, to newer fields such as economics, biosciences, and quantum mechanics, to name just a few.

Dimensional analysis and dimensional similarity can be applied to systems and control areas to provide solutions to known problems and to develop new approaches. In order to structure the potential applications of dimensional analysis and dimensional similarity to systems and control, the control problem is characterized by means of three elements, the physical system to be controlled, the model, and the controller.

The relations among the physical system, the model, and the controller are shown in Figure 1.4. The physical system and the model are related by modeling, identification, validation, and model reduction issues. Once a model is available, the model and the controller are related by control theory, of which one of the main objectives is to design proper controllers on the basis of mathematical models. Finally, the controller and the physical system are related by control engineering, aimed toward correct implementation and validation of control algorithms.

The conceptualization of control problems as presented in Figure 1.4 is useful to classify the current contributions of dimensional analysis to control problems and to envisage further applications. In this book we consider the following applications of dimensional analysis to control theory:

1. Identification and model validation:
   - Simple identification algorithms can be designed exploiting the model dimensionless representation.
   - The dimensionless representation is also useful in model order reduction for multivariable systems.

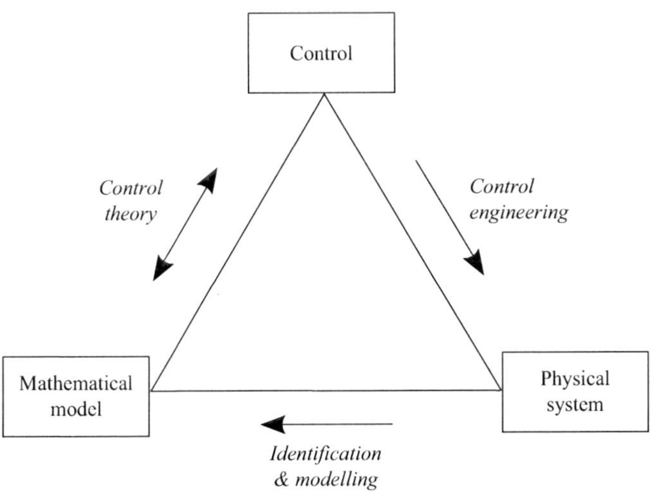

Figure 1.4 Elements and relations of control systems

2. Control theory:
   - The homogeneity property of proportional integral derivative (PID) tuning rules is shown to be a fundamental property that has influence on the type of functions used for deriving PID tuning rules.
   - The use of dimensionally similar reference models in model reference adaptive control with input constraints allows for the guarantee of stability while maintaining a dimensionally similar behavior at the output.
3. Control engineering:
   - The PID tuning rules comparison and selection are simplified by the homogeneity property of tuning rules and the application of a dimensionless framework.
   - The controller extrapolation from laboratory model experiments to prototypes can be readily performed by scaling.

1.4.1 Identification and model validation

The dimensionless representation of dynamical systems, both in the transfer function approach and in the state space representation, is presented in Chapter 3. Once a dimensionless representation is obtained, it is straightforward to characterize the dimensional similarity as shown in Chapter 4.

The application to system identification of the dimensionless representation permits to obtain simpler identification algorithms by exploiting limiting cases in the dimensionless numbers. An example is presented in Chapter 5, where a very simple graphical algorithm for identification of second order inverse response systems is presented. Other existing approaches in the literature need to perform integration or to solve a set of nonlinear equations. In the same way, the model order can be reduced by applying dimensional analysis. An application example of model order reduction of a multivariable system for decentralized control is also presented.

1.4.2 Control theory

In Chapter 6 the fundamental property of dimensional homogeneity is applied to PID tuning rules. Questions related to optimality and closed loop dimensional similarity are investigated. It follows that although the PID tuning rules homogeneity is not mandatory for correctness, it has implications that make homogeneity a natural choice.

Adaptive control also beneficiates from dimensional analysis. In Chapter 9 it is shown how the design of model reference adaptive controllers with a reference model that is no longer constant but changes in a dimensionally similar way yields a suitable control scheme in the presence of input saturation.

1.4.3 Control engineering

Dimensional analysis also has direct application to control engineering. The homogeneity of tuning rules has an important practical application in the derivation of procedures for PID tuning rules comparison as presented in Chapter 7. In fact, the existence of a huge set for PID tuning rules makes it difficult to choose which one is the most suited for the problem at hand. In the case of homogeneous tuning rules, it follows that the comparison may be performed in a dimensionless way, thus exploiting the benefits of parameter reduction and simplification of the limit cases.

Another application of dimensional similarity to control engineering is the use of laboratory-controlled models to design controllers that, by proper scaling, may be applied to prototypes. Chapter 8 presents the fundamental results for both, complete similarity and partial similarity. Partial similarity is of interest because to achieve complete similarity is costly and, sometimes, not possible. In this case it is shown how robustness properties are considered, enabling the use of experimental results to design controllers for prototypes.

1.5 Book contents

Chapter 2 The fundamental concepts, theorems, and results of dimensional analysis and dimensional similarity are introduced in this chapter. Although the chapter is self-contained, bibliographical references are presented to guide on further developments and applications of dimensional analysis to engineering problems. First the fundamental definitions of physical dimension, systems of units, and independence of dimensions are introduced. The fundamental Buckingham pi theorem of dimensional analysis is presented from a scalar viewpoint and from an algebraic approach. In this way results can be easily obtained for both the transfer function approach and the state space representation. Finally, the concepts and subtleties of dimensional similarity are explained.

Chapter 3 The main objective of this chapter is to obtain dimensionless representation of models given in transfer function form and state space form. The dimensionless form is advantageous because on the one hand the dimensionless representation has the minimum number of parameters required to define the dynamical model and, on the other hand, several distinct dimensional models may have the same dimensionless representation. As a result, being the dimensionless representation of a compressed representation of dynamic systems, it may be used as a measure of system complexity. The dimensionless representation of dynamical systems is the beginning for many of the applications of dimensional analysis and similarity to systems and control. In fact the dimensionless representation characterizes the similarity between dynamical systems.

Chapter 4 The dimensionless representation defines the concept of dimensional similarity between dynamical systems. The set of dimensional models that have the same dimensionless representation is the set of dimensionally similar models. In this way, the dimensionless representation characterizes the dimensionally similar models as the models that have the same dimensionless representation. The certification of dimensional similarity between systems allows the extrapolation of experimental results from model to prototype, and the application on novel adaptive control schemes in the presence of input saturation.

Chapter 5 The use of dimensionless models can be exploited in the identification and model order reduction procedures. The dimensionless representation makes it possible to point out limiting cases that may be exploited by identification algorithms and to reduce the order of models. A receipt is given in order to exploit the limiting cases, and it is shown how it can be used in the identification of second order inverse response systems and in the model reduction for decentralized control of multivariable processes.

Chapter 6 Dimensional analysis raises the question of the dimensional homogeneity of PID tuning rules. Whereas physical equations must be homogeneous for correctness, the PID tuning rules are just an algorithm that relates system parameters with controller parameters. The chapter presents the properties of homogeneous tuning rules and their implications on the process control. It follows that the homogeneity property of the tuning rules implies the mapping of dimensionally similar open loop systems into dimensionally similar closed loop systems, allowing for instance the development of frameworks for PID tuning rules comparison.

Chapter 7 Following with PID tuning rules, if they are homogeneous, it is possible to derive a comparative framework based on dimensionless numbers. It has the benefit of using the minimum required set of parameters for PID comparison, the dimensionless numbers, and it is possible to exploit the limiting values. The chapter presents a recipe to derive a suitable comparative framework and one example is presented.

Chapter 8 In the case of dimensional similarity experimental results from models can be generalized to the prototype. In this chapter the extrapolation of controllers designed on the basis of models to be used on prototypes is investigated. The case of complete similarity and the case of partial similarity are considered, where robust issues are considered for safer controller extrapolation.

Chapter 9 In this chapter we show how dimensional analysis can be used to design adaptive schemes. A novel model reference adaptive scheme is presented. The novelty of the approach lies on the fact that the reference model is no longer constant but dimensionally similar. In this way, although there are input constraints, the behavior is always similar to the original model. It implies that the dimensionless model properties are kept constant.

Chapter 2

Dimensional analysis and dimensional similarity

2.1 Physical quantities, units, and dimensions

2.1.1 Physical quantity

A physical quantity is a physical property or quality that can be quantified by measurement. A physical quantity is defined by the product of a numerical value and a unit of measurement, that is

$$Physical\ quantity = (Numerical\ value) \times (Unit) \tag{2.1}$$

For instance, when we refer to the fact that the length of a cable is 5 meters, we are stating the magnitude of a physical quantity.

2.1.2 Units

In order to obtain a physical quantity it is necessary to perform a measurement. The measurement is the comparison of the physical quantity with a standard, the unit of measurement. Following the cable length example, we could also say that the cable length is 5000 millimeters, the only difference being the units used to describe the physical quantity. Thus, in both cases we are referring to the same physical quantity (i.e. the cable length is the same), but the difference with the numerical values is due to the difference in the units used for measurement (e.g. meters vs millimeters).

2.1.3 Dimensions: fundamental and derived

The dimension is defined as a qualitative description of a sensory perception of a physical property or quality. Previous examples of physical quantities deal with the same dimension, the length dimension. We write the dimension of a physical quantity ϕ between square brackets, that is $[\phi]$. For example the quantity l_c that describes the length of the cable has length dimensions, that is $[l_c] = L$.

The dimensions can be classified into fundamental and derived. The fundamental dimensions are the ones arbitrarily chosen for the study of a physical phenomenon. For instance consider a mechanical phenomenon. The fundamental dimensions of the international system of units (SI) for the mechanical phenomenon are length L, mass M, and time T. Furthermore, the fundamental units are meter m, kilogram kg, and second s.

Derived quantities are obtained from fundamental ones according to physical laws. Correct derived quantities are obtained by multiplication and division of fundamental ones, being addition and subtraction forbidden. For example the velocity dimension is obtained by dividing the fundamental length dimension by the fundamental time dimension, that is $[v] = LT^{-1}$. The corresponding derived unit is meters per second ms^{-1}.

2.1.4 Arithmetic of dimensions

In this section we review the fundamentals of the arithmetic of dimensions. The results are quite straightforward and are presented without proof. The reader is referred to [2, p. 95] for a deeper treatment.

Product

The product of the dimensions of a set of n variables is the dimension of the product of these sets of n variables, that is if $v_1, v_2, \ldots, v_n$ is a set of n variables, then

$$[v_1][v_2], \ldots, [v_n] = [v_1, v_2, \ldots, v_n] \tag{2.2}$$

Power

The dimension of a power of a variable is the power of the dimension of that variable, that is

$$[v^n] = [v]^n \tag{2.3}$$

Quotient

The dimension of the quotient of nth-order differentials of variable v_1 with respect to variable v_2 is the quotient of dimensions of v_1 and v_2^n, that is

$$\left[\frac{d^n v_1}{dv_2^n}\right] = \frac{[v_1]}{[v_2^n]} \tag{2.4}$$

Integral

The dimension of a multiple integral is the product of the dimensions of the integrand and the dimensions of all the independent variables. That is, if

$$I = \int\int \cdots \int y(v_1, v_2, \ldots, v_n) dv_1 dv_2, \ldots, dv_n \tag{2.5}$$

then

$$[I] = [y(v_1, v_2, \ldots, v_n)][v_1][v_2], \ldots, [v_n] \tag{2.6}$$

2.2 Systems of units: dependence and independence of dimensions

2.2.1 System of units

A system of units consists of all the fundamental and derived dimensions, together with the units chosen for their measurement, for instance the centimeter-gram-second (CGS) system or the international system (SI). Table 2.1 presents the fundamental dimensions and their measurement units used in the SI. In Table 2.2 some of the SI-derived units are also presented.

Note that with reference to the dimensions length, mass, and time, both CGS and SI share the same physical properties, although the measurement units are different. A set of units that differ only in the measurement units but have the same dimensions is called a class of systems of units.

Table 2.1 Fundamental dimensions and units of SI

Physical quantity	**Symbol**	**Name of SI unit**	**Abbreviation**
Length	L	Meter	m
Mass	M	Kilogram	kg
Time	T	Second	s
Electric current	A	Ampere	A
Thermodynamic temperature	Θ	Kelvin	K
Amount of substance	M_l	Mole	mol
Luminous intensity	I	Candela	cd

Table 2.2 Derived SI units

Unit	**Symbol**	**Abbreviation**
Degree celsius	°C	K
Hertz	Hz	s^{-1}
Newton	N	$m \cdot kg \cdot s^{-2}$
Pascal	Pa	$N \cdot m^{-2}$
Joule	J	$N \cdot m$
Watt	W	$J \cdot s^{-1}$
Volt	V	$W \cdot A^{-1}$
Coulomb	C	$s \cdot A$
Ohm	Ω	$V \cdot A^{-1}$
Siemens	S	$A \cdot V^{-1}$
Farad	F	$C \cdot V^{-1}$
Weber	Wb	$V \cdot s$
Henry	H	$Wb \cdot A^{-1}$
Tesla	T	$Wb \cdot m^{-2}$

2.2.2 Monomial power law

The dimensions of a physical quantity are always a power law monomial. In this way, if we consider length, mass, and time dimensions, any derived dimension is always of the form:

$$[Quantity] = L^a M^b T^c \tag{2.7}$$

where a, b, and c are dimensionless coefficients. The result can be proven by using two systems of units of the same class, as shown in Reference 1 or by approximation theory [2, p. 133]. The result follows by the fact that all systems of units within a given class are equivalent.

2.2.3 Dependent and independent dimensions

Given a set of quantities $a_1, \ldots, a_k$, these quantities are said to have independent dimensions if none of these quantities have dimensions that can be represented in terms of a product of powers of the dimensions of the remaining quantities. That is there are no $p, \ldots, q, u, \ldots r$, such that

$$[a_j] = [a_1]^p, \ldots, [a_{j-1}]^q [a_{j+1}]^u, \ldots, [a_k]^r, \quad j = 1, \ldots, k \tag{2.8}$$

For example density ($[\rho] = ML^{-3}$), velocity ($[v] = LT^{-1}$), and force ($[f] = MLT^{-2}$) have independent dimensions. In fact there are no x and y that accomplish the following equations:

$$[\rho] = [v]^x [f]^y \tag{2.9}$$

$$LMT^{-2} = (ML^{-3})^x (LT^{-1})^y \tag{2.10}$$

because the set of equations

$$-3x + y = 1 \tag{2.11}$$

$$x = 1 \tag{2.12}$$

$$y = 2 \tag{2.13}$$

has no solution. The concept of independence of dimensions is crucial for the fundamental result of dimensional analysis, the Buckingham pi theorem.

2.3 Buckingham pi theorem

Consider a physical relation with N_v parameters involved, where the governed parameter a is a function of N_{v-1} governing parameters, that is

$$a = f(a_1, \ldots, a_{N_d}, a_{N_{d+1}}, \ldots, a_{N_{v-1}}) \tag{2.14}$$

The Buckingham pi theorem [1] states that if there is a set of N_d parameters $(a_1, \ldots, a_{N_d})$, which have independent dimensions then there are $N_v - N_d$ dimensionless parameters. One dimensionless number Π is related to the governed parameter a and Π_j $j \in [1, N_v - N_d - 1]$ are related to the $N_v - N_d - 1$ governing parameters, in such a way that the relation (2.14) can be stated as

$$\Pi = \Phi(\Pi_1, \ldots, \Pi_{N_v - N_d - 1}) \tag{2.15}$$

with Π the dimensionless number related with the governed parameter defined as

$$\Pi = \frac{a}{a_1^p, \ldots, a_{N_d}^r} \tag{2.16}$$

where the exponents $p, \ldots, r$ are chosen such that the parameter Π is dimensionless. Moreover, each one of the dimensionless numbers related with the governing parameters is given by

$$\Pi_j = \frac{a_{N_d+j}}{a_1^{p_j}, \ldots, a_{N_d}^{r_j}} \tag{2.17}$$

for $j \in [1, N_v - N_d - 1]$, where the exponents $p_j, \ldots, r_j$ are chosen such that the parameter Π_j is dimensionless. As a result there are $N_v - N_d - 1$ dimensionless numbers of the governing parameters (i.e. Π_j $j \in [1, N_v - N_d - 1]$), plus one dimensionless number related with the governed parameter a (i.e. Π). It then follows that the function f can be written in terms of a function of a smaller number of variables in the following form:

$$f(a_1, \ldots, a_{N_{v-1}}) = a_1^p, \ldots, a_{N_d}^r \Phi(\Pi_1, \ldots, \Pi_{N_v - N_d - 1}) \tag{2.18}$$

Example 2.1. Consider the capacitor charge through a resistance as presented in Figure 1.2. The capacitor voltage v_0 is described by the dimensional function $v_0(t) = f(t, v_i, r, c)$, where v_i is the input voltage, r is the resistance, c is the capacitance, and t is time. The dimensions of the involved variables are $[v_0] = [v_i] = V$, $[t] = T$, $[r] = V/A$, and $[c] = AT/V$, where V is voltage dimension, T is time dimension, and A is electrical current dimension. The number of variables is $N_v = 5$, whereas the number of variables with independent dimensions is $N_d = 3$, as a result there are $N_v - N_d = 2$ dimensionless numbers. A possible subset of the governing parameters with independent dimension is $\{v_i, r, c\}$. As a result the dimensionless governed parameter is given by

$$\Pi = \frac{v_0}{v_i} \tag{2.19}$$

and the dimensionless number of the governing parameter is

$$\Pi_1 = \frac{t}{rc} \tag{2.20}$$

so the capacitor charge can be written in dimensionless form as

$$v_0 = v_i \Phi\left(\frac{t}{rc}\right) \tag{2.21}$$

2.4 Matrix approach for finding the dimensionless numbers

In this section we present a structured approach in order to obtain the dimensionless numbers of any dimensional relation. The approach is based on matrix computations, so it is very convenient for computer implementation.

2.4.1 The dimensional matrix

The dimensional matrix [2, p. 134] is a matrix representation of the dimensions of the involved variables. The rows of the dimensional matrix are the dimensions, and the columns are the variables. The entries of the dimensional matrix are the exponents to which each particular dimension must be raised in the particular variable. For instance, given the dimensional matrix presented in Table 2.3, the dimension of variable v_5 is $d_1^5 d_2^2 d_3^1$.

Table 2.3 Dimensional matrix

	v_1	v_2	v_3	v_4	v_5
d_1	1	2	3	4	5
d_2	2	4	3	0	2
d_3	3	4	3	2	1

The dimensional matrix has N_d rows and N_v columns, where N_d is the number of dimensions required to describe the problem and N_v is the number of relevant variables necessary to describe the problem. It is assumed that they are not irrelevant variables. Furthermore, it is also assumed that $N_v \geq N_d$.

The dimensional matrix is partitioned into two submatrices, the M matrix and the N matrix. The M matrix is square, of order $N_d x N_d$, and non-singular. It is formed by the rightmost N_d columns of the dimensional matrix. The N matrix is formed by the first $N_v - N_d$ columns of the dimensional matrix. Note that the partition may be non-unique, although the fundamental requirement is that M matrix must be full rank. For instance, one possible partition of the dimensional matrix given in Table 2.3 is presented in Table 2.4. Note that the M matrix is full rank.

Conceptually, the variables related to the M matrix form the set of parameters with independent dimensions introduced in the Buckingham pi theorem. The

Table 2.4 Partition of the dimensional matrix

N matrix

1	2
2	2
3	4

M matrix

3	4	5
3	0	2
3	2	1

full rank requirement of M matrix reflects the dimension independence of the N_d variables. The variables related to the N matrix include the rest of variables.

2.4.2 The dimensional set

The dimensional set is a matrix representation that yields the dimensionless group number in a compact and economic way. The dimensional set is formed by four matrices, as presented in Table 2.5. N and M matrices are the dimensional matrix previously defined. P matrix is the identity, of size $(N_v - N_d)x(N_v - N_d)$. The O matrix is given by $O = -P(M^{-1}N)^T$, and it provides the exponents to which the $N_v - N_d$ dimensions must be raised to obtain the dimensionless numbers.

Table 2.5 Dimensional set

	Physical variables	
Dimensions	*N* matrix	*M* matrix
Dimensionless var.	*P* matrix	*O* matrix

Example 2.2. Consider again the capacitor charge through a resistance as presented in Figure 1.2. The dimensional matrix is given in Table 2.6.

Table 2.6 Capacitor charge circuit dimensional matrix

	v_0	t	v_i	r	c
V	1	0	1	1	−1
A	0	0	0	−1	1
T	0	1	0	0	1

The dimensional matrix is partitioned as presented in Table 2.7, by choosing v_i, r, and c as variables with independent dimensions.

Table 2.7 N and M matrices for the capacitor charge circuit dimensional matrix

	v_0	t	v_i	r	c
V	1	0	1	1	−1
A	0	0	0	−1	1
T	0	1	0	0	1

We recall that the partition is not unique. In fact v_0 could be used instead of v_i in the M matrix. Even the set $\{t, v_0, r\}$ defines an M matrix that is of full dimension.

The resulting dimensional set is shown in Table 2.8. The dimensional numbers are given by matrices P and O. From the exponents of each variable we have

$$\Pi = \frac{v_0}{v_i} \tag{2.22}$$

and

$$\Pi_1 = \frac{t}{rc} \tag{2.23}$$

that is the same result as in the previous example.

2.5 Dimensional similarity

The Buckingham pi theorem states that the behavior of a physical system may be described in dimensionless form and is completely defined by a set of dimensionless numbers. In this way, if two systems have the same dimensionless numbers, their behavior must be correlated in some way, and the systems are dimensionally similar. Whenever two systems are dimensionally similar, it is possible to extrapolate the results in one system to the other. This is the fundamental idea of dimensional modeling.

Table 2.8 Capacitor charge circuit dimensional set

	v_0	t	v_i	r	c
V	1	0	1	1	−1
A	0	0	0	−1	1
T	0	1	0	0	1
Π	1	0	−1	0	0
Π_1	0	1	0	−1	−1

In order to define dimensional similarity we consider two systems, the prototype and the model, expressed by means of dimensionless numbers as presented in the preceding section, that is

$$\Pi_p = \Phi(\Pi_{p1}, \ldots, \Pi_{p(N_v - N_d - 1)}) \tag{2.24}$$

$$\Pi_m = \Phi(\Pi_{m1}, \ldots, \Pi_{m(N_v - N_d - 1)}) \tag{2.25}$$

Conceptually the prototype can be understood as the system to be designed while the model is a facility to be used for investigation and to extract information to be used in the prototype.

Definition 2.1 (Szirtes and Rózsa [2], p. 473). Two systems (prototype and model) are dimensionally similar if all the dimensionless variables describing the systems are identical in construction and in magnitude, that is

$$\begin{aligned} \Pi_p &= \Pi_m \\ \Pi_{p1} &= \Pi_{m1} \\ &\vdots \\ \Pi_{p(N_v - N_d - 1)} &= \Pi_{m(N_v - N_d - 1)} \end{aligned} \tag{2.26}$$

However, previous definition is redundant as Theorem 2.1 shows.

Theorem 2.1 (Szirtes and Rózsa [2], p. 473). *If in two systems (prototype and model) all the corresponding pairs of dimensionless variables containing only governing variables are identical in construction and in magnitude, that is*

$$\begin{aligned} \Pi_{p1} &= \Pi_{m1} \\ &\vdots \\ \Pi_{p(N_v - N_d - 1)} &= \Pi_{m(N_v - N_d - 1)} \end{aligned} \tag{2.27}$$

then the dimensionless variables forming the remaining pair (containing the governed variable, that is ($\{\Pi_p = \Pi_m\}$) *are also identical.*

The result easily follows from (2.24). The dimensional similarity definition between systems is given by the following theorem.

Theorem 2.2. *Assume that there is only one governed variable. Furthermore, the governed variable appears in only one of the dimensionless variables, that is the governed variable appears in the N matrix of the dimensional set. Then, if in two systems (prototype and model) all the corresponding pairs of dimensionless variables* *containing only governing variables are identical in construction and magnitude, that is*

$$\begin{aligned} \Pi_{p1} &= \Pi_{m1} \\ &\vdots \\ \Pi_{p(N_v-N_d-1)} &= \Pi_{m(N_v-N_d-1)} \end{aligned} \tag{2.28}$$

then the prototype and model are dimensionally similar.

Finally, we define the concept of partial similarity.

Definition 2.2. If in two systems (prototype and model) some dimensionless numbers related to the governing variables are not equal, then the systems are partially dimensionally similar.

Example 2.3. In the previous capacitor charge example, the dimensionless relation obtained was

$$\Pi = \Phi(\Pi_1) \tag{2.29}$$

with $\Pi = v_0/v_i$ the governed variable dimensionless number and $\Pi_1 = t/rc$ the governing variable dimensionless number. First note that there is only one governed variable (i.e. v_0) and that the governed variable appears in only one of the dimensionless variables (i.e. $\Pi = v_0/v_i$). In this case both assumptions are fulfilled and two systems with identical Π_1 dimensionless numbers are dimensionally similar, that is $\Pi_{p1} = \Pi_{m1}$ with $\Pi_{p1} = t_p/r_pc_p$ and $\Pi_{m1} = t_m/r_mc_m$.

2.5.1 Scale factors

Definition 2.3 (Szirtes and Rózsa [2], p. 479). The scale factor with respect to a particular physical variable is the quotient of the magnitudes of that variable for the prototype and its model.

It is important to note that there is not a unique scale factor between a model and a prototype, but for each physical variable there is one scaling factor.

Example 2.4. Consider the variables involved in the $\Pi_1 = t/rc$ dimensionless number of the capacitor charge example. The dimensional variables involved are t, r, and c. The scaling factors related with the previous dimensional variables are the time scaling factor $S_t = t_p/t_m$, the resistance scaling factor $S_r = r_p/r_m$, and the capacitance scaling factor $S_c = c_p/c_m$.

2.5.2 Model law

A model law is a set of relations among scaling factors that are derived from dimensional similarity constraints.

Example 2.5. In the capacitor charge example, for dimensional similarity between model and prototype, it must follow that $\Pi_{p1} = \Pi_{m1}$, that is

$$\frac{t_p}{r_p c_p} = \frac{t_m}{r_m c_m} \tag{2.30}$$

which can be rearranged as

$$\frac{t_p}{t_m} = \frac{r_p c_p}{r_m c_m} \tag{2.31}$$

and by definition of scaling factors it yields

$$S_t = S_r S_c \tag{2.32}$$

which is the model law that must follow any capacitor charge circuit in order to be dimensionally similar.

2.6 Exercises

1. Given a right triangle and considering the area a as the governed variable and the hypotenuse h and the hypotenuse angle α as the governing variables:
   a) What is the physical dimension and units of the hypotenuse angle α?
   b) Can α appear in the M matrix?
   c) Find one dimensional set and derive the dimensionless numbers.
   d) Derive the dimensionless relation.
2. In the ceramic tile glazing process, the tile is covered by a uniform layer of glazing material. The tile passes under a continuous cascade of liquid glaze. The variable to be controlled is the quantity (mass) of glaze deposited per tile. It is assumed that the glaze mass deposited m_l with units $[m_l] = \mathrm{kg/m}$ depends on the glaze cascade flow q, $[q] = \mathrm{m^3/s}$, the tile velocity v, $[v] = \mathrm{m/s}$, and the glaze density ρ, $[\rho] = \mathrm{kg/m^3}$.
   a) Classify the variables as governing or governed variables.
   b) Derive a dimensionless relation among the previous variables by direct application of the Buckingham pi theorem.
   c) Find one dimensional set and derive the dimensionless numbers that define the ceramic tile glazing process.
   d) Obtain the model law between a model and a prototype for dimensional similarity.

3. In the previous ceramic tile glazing process, it is important to stress the difference between the glaze cascade flow q and the deposited flow in the tile q_d. It is assumed that the difference depends on the length of the tile l_t and the characteristic length of the cascade l_c.
   a) Construct a dimensional set if the governed variable is q.
   b) If a model must be constructed, find the model law.
   c) For dimensional similarity, is there any constraint between the cascade flow q and the tile length l_t?

References

[1] G.I. Barenblatt. *Dimensional Analysis*. Gordon and Breach Science Publishers, 1987

[2] T. Szirtes and P. Rózsa. *Applied Dimensional Analysis and Modeling*. Butterworth-Heinemann, 2007

Chapter 3
Dynamical systems: dimensionless representation

3.1 Introduction

The dimensionless representation minimizes the number of parameters necessary for the system description, and may provide further knowledge of the parameters influence on the dynamical response. Furthermore, as proposed in Reference 7, the number of dimensionless numbers is a measure of the complexity of the problem at hand. Henceforth, the dimensionless representation may provide a system complexity characterization that is different from mere system order. In fact, a small number of dimensionless numbers imply that the dynamical system can be characterized with a small number of experiments. For instance, in fluid dynamics problems dominated by gravitational forces, only the Froud number is of relevance, so Euler and Reynolds dimensionless numbers are not considered. On the contrary, when the system dynamics depends on a large amount of relevant dimensionless numbers, the characterization of the dynamical system requires a great number of experiments. For example consider the plasma confinement problem. In this case all the hydrodynamic, electromagnetic, and nuclear dimensionless numbers are relevant in the plasma confinement, making the dynamical behavior very complex [7].

Another property of the dimensionless representation is that it is the fingerprint of dimensional similarity. In this way two dimensionally similar systems have the same dimensionless representation. Consequently, the analysis and design results over dimensionless models are applicable not only to one dynamical model, but also to the infinite family of all dimensionally similar models.

In this chapter we show how to obtain the dimensionless form of dynamical models represented in transfer function and in state space. The similarities and implications of dimensionless models are stated and compared with other dynamical system transformations such as coordinate transformation and signal scaling. Finally, we perform the dimensionless transformation of discrete time systems, showing its differences with respect to continuous time models. The dimensionless representation of dynamical systems forms the backbone for the applications of dimensional analysis to control theory presented in the rest of the book.

3.2 Transfer function dimensionless representation

The transfer function dimensionless representation is based on direct application of the Buckingham pi theorem. First, it is worth to remember that physical parameters such as mass, length, capacitance, etc., are dumped into transfer function parameters like gain, time constants, and time delay.

Let's consider the application of the Buckingham pi theorem to a transfer function to find a dimensionless representation. The Buckingham pi theorem requires first, to find a set of transfer function parameters with independent dimensions, second, to find the dimensionless numbers, and finally to express the transfer function in dimensionless form.

3.2.1 Transfer function parameters dimensions

Prior to find a set of transfer function parameters with independent dimensions, the physical dimensions of transfer function parameters must be determined [1]. The dimensions of the transfer function parameters are given by the following lemma.

Lemma 3.1. *Consider the generic transfer function $G(s)$ in pole-zero form*

$$G(s) = k\frac{\prod_{j=1}^{m}(\tau_{cj}s+1)}{s^l\prod_{i=1}^{n}(\tau_{pi}s+1)}e^{-hs} \tag{3.1}$$

with transfer function parameters $\theta_G = k, h, \tau_{p1}, \ldots, \tau_{pn}, \tau_{c1}, \ldots, \tau_{cm}, l$. The relative system order is $n + l - m \geq 0$. K is the gain dimension and T the time dimension. The dimensions of the transfer function parameters are $[k] = \mathrm{K}/(\mathrm{T}^l)$, $[\tau_{pi}] = [\tau_{cj}] = \mathrm{T}$ for $i = 1, \ldots, n$ and $j = 1, \ldots, m$, $[h] = \mathrm{T}$, and $[l] = 1$, where 1 stands for dimensionless quantity.

Proof. First we show that $\prod_{i=1}^{n}(\tau_{pi}s+1)$ is dimensionless, thus $[\prod_{i=1}^{n}(\tau_{pi}s+1)] = 1$. The dimension of τ_{pi} for $i = 1, \ldots, n$ is time dimension $[\tau_{pi}] = \mathrm{T}$. On the other hand, the dimension of the s operator is $[s] = \mathrm{T}^{-1}$, as can be seen either in the time domain, where the s operator is transformed to operator d/dt, or in the frequency domain, where s is transformed to $j\omega$. Consequently, each product $\tau_{pi}s$ yields a dimensionless number. The same reasoning applies to $\prod_{j=1}^{m}(\tau_{cj}s+1)$ and e^{-hs}. The parameter l is also dimensionless because exponents of transcendental functions must be dimensionless for homogeneity [6, p. 104]. Consequently, $[G(s)] = [ks^{-l}]$ or equivalently, $[k] = [G(s)][(s^l)]$ that yields $[k] = \mathrm{K}/(\mathrm{T}^l)$. □

As a result, only two physical dimensions are required to label the physical dimensions of transfer function parameters, the gain dimension K and the time dimension T, thus $N_d = 2$.

3.2.2 Transfer function parameters with independent dimensions

Once the fundamental dimensional set is established and the dimensions of the transfer function parameters determined, we can construct the dimensional matrix of transfer function parameters, as presented in Table 3.1. It can be seen that all the parameters can be divided into three sets, the parameters with gain and/or time dimension $\{G, k\}$, the parameters with time dimension only $\{\tau_{p1}, \ldots, \tau_{pn}, \tau_{c1}, \ldots, \tau_{cm}, h, s\}$, and the dimensionless parameter $\{l\}$.

Table 3.1 Dimensional matrix of transfer function parameters

	G	k	τ_{p1}	$\ldots$	τ_{pn}	τ_{c1}	$\ldots$	τ_{cm}	h	s	l
K	1	1	0	$\ldots$	0	0	$\ldots$	0	0	0	0
T	0	$-l$	1	$\ldots$	1	1	$\ldots$	1	1	-1	0

By choosing k and τ_{p1} as the set of parameters with independent dimensions, the dimensional matrix can be partitioned as shown in Table 3.2. It can be seen that k and τ_{p1} are independent because matrix M has full rank, even if $l = 0$.

Note that choosing k and τ_{p1} as parameters with independent dimension is arbitrary. However, the parameter l, which is already dimensionless, cannot be a column of the M matrix, because in this case matrix M would be singular. In the same way, any two parameters that depend on the time dimension have always a singular M matrix, thus they have not independent dimensions. Finally, it is possible to consider G and k as a set of parameters with independent dimensions because the M matrix would be in this case

$$M = \begin{bmatrix} 1 & 0 \\ 0 & -l \end{bmatrix} \tag{3.2}$$

and for $l \neq 0$ it has full rank. However, l is the number of transfer function integrators, and in the case of no integrators $l = 0$, hence the M matrix becomes singular. Consequently in the general case it is not possible to choose the pair G and k as parameters with independent dimensions.

Table 3.2 Partition of the dimensional matrix of transfer function parameters

	G	τ_{p2}	$\ldots$	τ_{pn}	τ_{c1}	$\ldots$	τ_{cm}	h	s	l	k	τ_{p1}
K	1	0	$\ldots$	0	0	$\ldots$	0	0	0	0	1	0
T	0	1	$\ldots$	1	1	$\ldots$	1	1	-1	0	$-l$	1

3.2.3 Transfer function dimensionless numbers

We have established that the number of independent dimensions is 2, that is $N_d = 2$, and a convenient set of transfer function parameters with independent dimensions is $\{k,\ \tau_{p1}\}$. The next step, following the Buckingham pi theorem, is to obtain the $N_v - N_d$ dimensionless numbers. Taking as starting point the dimensional matrix partition of Table 3.2, we construct the dimensional set presented in Table 3.3, with matrix P an identity matrix of size $(N_v - N_d)x(N_v - N_d)$, and matrix $O = -P(M^{-1}N)^T$.

Each row of the P and O matrices provide a dimensionless number Π_i, $i = 1, \ldots, N_v - 2$, as can be seen in Table 3.2. For instance Π_1 is given by

$$\Pi_1 = \frac{G(s)}{k\tau_{p1}^{l}} \tag{3.3}$$

Note that, by P matrix being the identity, there is one dimensionless number for each one of the transfer function parameters with dependent dimensions $\{G, \tau_{p2}, \ldots, \tau_{pn}, \tau_{c1}, \ldots, \tau_{cm}, h, s, l\}$. As a result, for a generic transfer function with integrators and time delay, the quantity of dimensionless numbers is equal to $N_v - N_d = n + m + 4$, where n is the number of poles and m the number of zeroes. The reduction of transfer function dimensional parameters N_d is limited to $N_d = 2$.

The dimensionless transfer function parameters can be interpreted as follows:

- The dimensionless number $\Pi_1 = \frac{G(s)}{k\tau_{p1}^{l}}$ is the dimensionless transfer function.
- The dimensionless numbers $\Pi_i = \tau_{pi}/\tau_{p1}$, $i = 2, \ldots, n$ and $\Pi_{n+j} = \tau_{cj}/\tau_{p1}$, $j = 1, \ldots, m$, are the dimensionless time constants.
- The dimensionless number $\Pi_{N_v-4} = h/\tau_{p1}$ is the dimensionless time delay.
- The dimensionless number $\Pi_{N_v-3} = s\tau_{p1}$ is the dimensionless Laplace variable.
- The dimensionless number $\Pi_{N_v-2} = l$ is the number of integrators, which is already dimensionless.

Table 3.3 Transfer function dimensional set

	G	τ_{p2}	$\ldots$	τ_{pn}	τ_{c1}	$\ldots$	τ_{cm}	h	s	l	k	τ_{p1}
K	1	0		0	0		0	0	0	0	1	0
T	0	1		1	1		1	1	-1	0	$-l$	1
Π_1	1	0		0	0		0	0	0	0	-1	$-l$
Π_2	0	1		0	0		0	0	0	0	0	-1
$\vdots$											$\vdots$	$\vdots$
Π_n	0	0		1	0		0	0	0	0	0	-1
Π_{n+1}	0	0		0	1		0	0	0	0	0	-1
$\vdots$											$\vdots$	$\vdots$
Π_{n+m}	0	0		0	0		1	0	0	0	0	-1
Π_{N_v-4}	0	0		0	0		0	1	0	0	0	-1
Π_{N_v-3}	0	0		0	0		0	0	1	0	0	1
Π_{N_v-2}	0	0		0	0		0	0	0	1	0	0

3.2.4 Dimensionless transfer function

Once the transfer function dimensionless numbers are determined we provide a straightforward methodology to represent a transfer function in a dimensionless form. The procedure is as follows:

(i) Choose the transfer function dominant time constant τ_{p1}.
(ii) Divide each transfer function parameter with time dimension by τ_{p1}.
(iii) Perform the following variable substitution $s \rightarrow \bar{s} := \tau_{p1}s$. We obtain $G(\bar{s})$.
(iv) Divide the resulting transfer function $G(\bar{s})$ by $k\tau_{p1}^{l}$ to obtain the dimensionless transfer function $\bar{G}(\bar{s})$.

Steps (ii) and (iii) leave the dimensional transfer function unaltered, that is $G(s) = G(\bar{s})$. $G(\bar{s})$ has dimension equal to $[G(\bar{s})] = \mathrm{KT}^{l}$, while transfer function $\bar{G}(\bar{s})$ is dimensionless, that is $[\bar{G}(\bar{s})] = 1$.

Example 3.1. Consider the second order system with time delay and $\tau_1 \geq \tau_2$

$$G(s) = \frac{ke^{-hs}}{(\tau_1 s + 1)(\tau_2 s + 1)} \tag{3.4}$$

The quantity of variables is $N_v = 5$, the quantity of dimensions is $N_d = 2$, and the expected quantity of dimensionless numbers is $N_v - N_d = 3$. First, following step (i) we choose the dominant time constant τ_1 as normalising variable. After steps (ii) and (iii) we have

$$G(\bar{s}) = \frac{ke^{-\bar{h}\bar{s}}}{(\bar{s} + 1)(\bar{\tau}_2 \bar{s} + 1)} \tag{3.5}$$

with $\bar{h} = \frac{h}{\tau_1}$, $\bar{\tau_2} = \frac{\tau_2}{\tau_1}$, and $\bar{s} = \tau_1 s$. Finally, step (iv) provides the dimensionless transfer function

$$\bar{G}(\bar{s}) = \frac{G(\bar{s})}{k} = \frac{e^{-\bar{h}\bar{s}}}{(\bar{s} + 1)(\bar{\tau}_2 \bar{s} + 1)} \tag{3.6}$$

3.3 State space dimensionless representation

The development of modern control theory had its roots in the state space representation of dynamical systems and the concept of system state space [3]. The standard linear time invariant (LTI) state space representation is given by

$$\dot{x}(t) = Ax(t) + Bu(t) \tag{3.7}$$

$$y(t) = Cx(t) + Du(t) \tag{3.8}$$

with state vector $x(t) \in \mathbb{R}^n$, input vector $u(t) \in \mathbb{R}^p$, output vector $y(t) \in \mathbb{R}^q$, dynamic matrix $A \in \mathbb{R}^{n\times n}$, input matrix $B \in \mathbb{R}^{n\times p}$, output matrix $C \in \mathbb{R}^{q\times n}$, and matrix $D \in \mathbb{R}^{q\times p}$.

The first step, following Buckingham pi theorem, would be to determine the dimensions of matrices A, B, C, and D of the state space representation. Although in Reference 4 the dimensions of matrix A are determined, the determination of the matrices dimensions is not as straightforward as in the transfer function case because the matrix entries are combinations of physical parameters.

The approach taken to find the dimensionless representation of the state space representation ((3.7)–(3.8)) is based on system transformation, as presented in Reference 2. In this way, first we construct the following dimensionless groups:

$$x(t) = N_x\bar{x}(\bar{t}) \tag{3.9}$$

$$u(t) = N_u\bar{u}(\bar{t}) \tag{3.10}$$

$$y(t) = N_y\bar{y}(\bar{t}) \tag{3.11}$$

$$t = N_t\bar{t} \tag{3.12}$$

where $N_x \in \mathbb{R}^{n\times n}$, $N_u \in \mathbb{R}^{p\times p}$, $N_y \in \mathbb{R}^{q\times q}$, $N_t \in \mathbb{R}$, and $\bar{x}(\bar{t})$ is the dimensionless state space, $\bar{u}(\bar{t})$ the dimensionless input, $\bar{y}(\bar{t})$ the dimensionless output, and $\bar{t}$ the dimensionless time.

To obtain the dimensionless representation first note that

$$\begin{aligned}\frac{dx(t)}{dt} &= \frac{d}{dt}(N_x\bar{x}(\bar{t}))\\ &= N_x\frac{d\bar{x}(\bar{t})}{dt}\\ &= N_x\frac{d\bar{x}(\bar{t})}{d\bar{t}}\frac{d\bar{t}}{dt}\\ &= N_x\dot{\bar{x}}N_t^{-1}\end{aligned} \tag{3.13}$$

and by direct substitution of ((3.9)–(3.13)) into ((3.7)–(3.8)), we have

$$N_xN_t^{-1}\dot{\bar{x}}(t) = AN_x\bar{x}(t) + BN_u\bar{u}(t) \tag{3.14}$$

$$N_y\bar{y}(t) = CN_x\bar{x}(t) + DN_u\bar{u}(t) \tag{3.15}$$

which can be rewritten in dimensionless form as

$$\dot{\bar{x}}(\bar{t}) = \bar{A}\bar{x}(\bar{t}) + \bar{B}\bar{u}(\bar{t}) \tag{3.16}$$

$$\bar{y}(\bar{t}) = \bar{C}\bar{x}(\bar{t}) + \bar{D}\bar{u}(\bar{t}) \tag{3.17}$$

with dimensionless matrices

$$\begin{aligned}\bar{A} &= N_tN_x^{-1}AN_x\\ \bar{B} &= N_tN_x^{-1}BN_u\\ \bar{C} &= N_y^{-1}CN_x\\ \bar{D} &= N_y^{-1}DN_u\end{aligned}$$

Example 3.2. Consider the state space representation of an inverted pendulum given by

$$\begin{bmatrix} \ddot{\theta} \\ \dot{\theta} \end{bmatrix} = \begin{bmatrix} -\frac{c}{i} & \frac{mgl - k}{i} \\ 1 & 0 \end{bmatrix} \begin{bmatrix} \dot{\theta} \\ \theta \end{bmatrix} + \begin{bmatrix} \frac{1}{i} \\ 0 \end{bmatrix} u \tag{3.18}$$

$$y = \begin{bmatrix} 0 & 1 \end{bmatrix} \begin{bmatrix} \dot{\theta} \\ \theta \end{bmatrix} \tag{3.19}$$

with θ the angular position, $\dot{\theta}$ the angular velocity, u the input torque, g the gravity acceleration, l the pendulum length, k the angular spring stiffness, c the friction constant, m the pendulum mass, and i the pendulum inertia.

From the dimensional set of Table 3.4, we can obtain the dimensionless signals as

Table 3.4 Inverted pendulum dimensional set

	θ	$\dot{\theta}$	u	t	g	l	c	i	m	k
M	0	0	1	0	0	0	1	1	1	1
L	0	0	2	0	1	1	2	2	0	2
T	0	−1	−2	1	−2	0	−1	0	0	−2
$\bar{\theta}$	1	0	0	0	0	0	0	0	0	0
$\bar{\dot{\theta}}$	0	1	0	0	0	0	0	1/2	0	−1/2
$\bar{u}$	0	0	1	0	0	0	0	0	0	−1
$\bar{t}$	0	0	0	1	0	0	0	−1/2	0	1/2
$\bar{g}$	0	0	0	0	1	0	0	1/2	1/2	−1
$\bar{l}$	0	0	0	0	0	1	0	−1/2	1/2	0
$\bar{c}$	0	0	0	0	0	0	1	−1/2	0	−1/2

$$\begin{aligned} \bar{\theta} &= \theta \\ \bar{\dot{\theta}} &= \dot{\theta}\sqrt{\frac{i}{k}} \\ \bar{u} &= u\frac{1}{k} \\ \bar{t} &= t\sqrt{\frac{k}{i}} \end{aligned}$$

The resulting dimensionless matrices are thus easily obtained as

$$\begin{aligned} N_x &= \begin{bmatrix} \sqrt{\frac{k}{i}} & 0 \\ 0 & 1 \end{bmatrix} \\ N_u &= \begin{bmatrix} k \end{bmatrix} \end{aligned}$$

$$N_y = [1]$$
$$N_t = \left[\sqrt{\frac{i}{k}}\right]$$

That finally yields the dimensionless representation of the state space inverted pendulum

$$\begin{bmatrix} \ddot{\bar{\theta}} \\ \dot{\bar{\theta}} \end{bmatrix} = \begin{bmatrix} -\bar{c} & \bar{g}\bar{l} - 1 \\ 1 & 0 \end{bmatrix} \begin{bmatrix} \dot{\bar{\theta}} \\ \bar{\theta} \end{bmatrix} + \begin{bmatrix} 1 \\ 0 \end{bmatrix} \bar{u} \tag{3.20}$$

$$\bar{y} = \begin{bmatrix} 0 & 1 \end{bmatrix} \begin{bmatrix} \dot{\bar{\theta}} \\ \bar{\theta} \end{bmatrix} \tag{3.21}$$

with

$$\bar{g} = g\frac{\sqrt{im}}{k}$$
$$\bar{l} = l\sqrt{\frac{m}{i}}$$
$$\bar{c} = c\sqrt{\frac{1}{ik}}$$

Note that all the dimensionless variables involved $\{\bar{\theta}, \dot{\bar{\theta}}, \bar{u}, \bar{t}, \bar{g}, \bar{l}, \bar{c}\}$ are obtained as the rows of the lower submatrices in the dimensional set (Table 3.4).

3.3.1 Interpretation of the state space dimensionless transformation

The state space model ((3.7)–(3.8)) is given in dimensionless form by the following transformation:

$$\bar{A} = N_t N_x^{-1} A N_x \tag{3.22}$$
$$\bar{B} = N_t N_x^{-1} B N_u \tag{3.23}$$
$$\bar{C} = N_y^{-1} C N_x \tag{3.24}$$
$$\bar{D} = N_y^{-1} D N_u \tag{3.25}$$

The transformations to obtain a dimensionless model can be interpreted in terms of a coordinate transformation, a signal scaling, and a time scaling. The state space vector x of a state space model can be transformed by a nonsingular matrix T, giving a new state space representation x_T, that is $x = Tx_T$ [3]. This transformation is also known as similar transformation. However, to avoid confusion with dimensional analysis similarity concepts, we will refer it as coordinate transformation.

The coordinate transformation provides a new state space representation given by matrices:

$$A_T = T^{-1}AT \tag{3.26}$$

$$B_T = T^{-1}B \tag{3.27}$$

$$C_T = CT \tag{3.28}$$

$$D_T = D \tag{3.29}$$

Another system transformation is the signal scaling that is motivated by a normalization of signal values. In this way, the input and output signals are scaled as $y = D_y y_s$ and $u = D_u u_s$, with D_y and D_u diagonal matrices that include the inverse of the maximum expected value of y and u, respectively [5]. The state space matrices after a signal scaling are given by

$$A_S = A \tag{3.30}$$

$$B_S = BD_u \tag{3.31}$$

$$C_S = D_y^{-1}C \tag{3.32}$$

$$D_S = D_y^{-1}DD_u \tag{3.33}$$

By comparison of (3.22)–(3.25) with (3.26)–(3.29) and (3.30)–(3.33), it can be seen that the state space dimensionless transformation includes both a coordinate transformation and a signal transformation. Moreover, the dimensionless transformation also includes a time scaling by means of factor N_t. Note that neither coordinate transformation nor signal scaling incorporates a timescaling.

Certain system properties are invariant under coordinate transformation and signal scaling, for instance the characteristic polynomial, the system output, and the system transfer function, among others. In Table 3.5 it can be seen how these properties are invariant under coordinate transformation and signal scaling. However, under dimensionless transformation, due to time scaling, we see that the characteristic polynomial is scaled by the time factor N_t, that is $|N_t||\bar{s}I - A|$, with $\bar{s} = s/N_t$. In the same way, the output is scaled by the time variable and given by

Table 3.5 Characteristic polynomial (Char. Pol), System output (Output), and transfer function (T.F.) for coordinate transformation, signal scaling, and dimensionless transformation. Note that $\bar{s} = s/N_t$ and $t = N_t\bar{t}$

	Coordinate transformation	**Signal scaling**	**Dimensionless transformation**
Char. Pol.	$\|sI - A\|$	$\|sI - A\|$	$\|N_t\|\|\bar{s}I - A\|$
Output	$y(t) = Cx(t) + Du(t)$	$y(t) = Cx(t) + Du(t)$	$y(\bar{t}) = Cx(\bar{t}) + Du(\bar{t})$
T.F.	$C(sI - A)^{-1}B + D$	$C(sI - A)^{-1}B + D$	$N_t^2(C(\bar{s}I - A)^{-1}B + D)$

$y(\bar{t}) = Cx(\bar{t}) + Du(\bar{t})$, with $t = N_t\bar{t}$. Finally, the transfer function is also time scaled by $N_t^2(C(\bar{s}I - A)^{-1}B + D)$, with $\bar{s} = s/N_t$.

3.4 Comparison between transfer function and state space dimensionless representation

The transfer function dimensionless representation is obtained from transfer function parameters such as poles and zeros time constants, gain, time delay, and number of integrators. On the contrary, the state space dimensionless representation is obtained from physical parameters such as capacitance and electrical resistance. The physical parameters of state space representation are dumped into transfer function parameters. In this way the dimensionless representation of transfer function loses physical insight. However, the obtention of dimensionless representation of transfer functions is straightforward and of great generality. In fact, the procedure for dimensionless representation of transfer functions is not dependent on the physical problem at hand.

The use of the transfer function representation or the state space representation in control theory depends on the type of problems to be solved. For instance, model reference adaptive control and process control are traditionally tackled under the transfer function formalism, whereas modern control theory concepts as controllability and observability arise in state space models. Their differences are shown by the following example.

Example 3.3. Consider the mass-spring-damper dynamical system of Figure 3.1 with initial conditions $x(0) = x_0$, $\dot{x}(0) = 0$, input $F(t)$, and output $x(t)$. The output is given by

$$X(s) = \frac{1}{ms^2 + cs + k}F(s) + \frac{sm + c}{ms^2 + cs + k}x_0 \tag{3.34}$$

that is represented in gain and time constants form as

$$X(s) = \frac{k_1}{(\tau_1 s + 1)(\tau_2 s + 1)}F(s) + \frac{k_2(\tau_3 s + 1)}{(\tau_1 s + 1)(\tau_2 s + 1)} \tag{3.35}$$

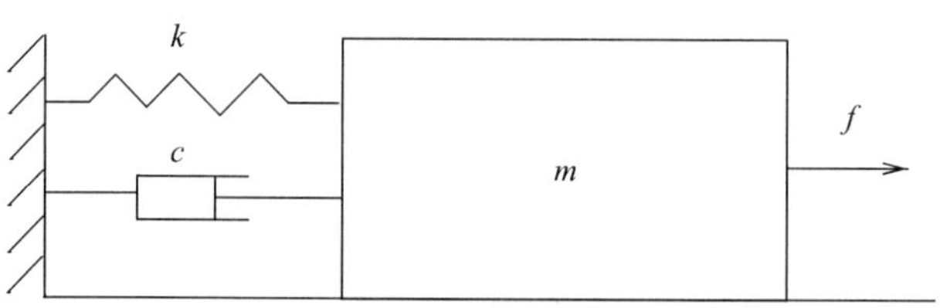

Figure 3.1 Mass-spring-damper dynamical system

with transfer function parameters

$$k_1 = \frac{1}{k} \tag{3.36}$$

$$k_2 = \frac{cx_0}{k} \tag{3.37}$$

$$\tau_1 = \frac{2m}{(c + \sqrt{(c^2 - 4km)})} \tag{3.38}$$

$$\tau_2 = \frac{2m}{(c - \sqrt{(c^2 - 4km)})} \tag{3.39}$$

$$\tau_3 = \frac{m}{c} \tag{3.40}$$

The dimensionless transfer function is easily obtained as

$$\frac{\bar{X}_F(s)}{F(s)} := \frac{X_F(s)}{k_1 F(s)} = \frac{1}{(\bar{s}+1)(\bar{\tau}_2\bar{s}+1)} \tag{3.41}$$

with $\bar{s} = \tau_1 s$ and $\bar{\tau}_2 = \tau_2/\tau_1$. The response due to initial conditions in dimensionless form is

$$\bar{X}_{x0}(s) := \frac{X_{x0}(s)}{k_2} = \frac{(\bar{\tau}_3\bar{s}+1)}{(\bar{s}+1)(\bar{\tau}_2\bar{s}+1)} \tag{3.42}$$

with $\bar{\tau}_3 = \tau_3/\tau_1$. The dimensionless system output is given by

$$\bar{X}(s) = \bar{X}_F(s) + \bar{X}_{x0}(s) \tag{3.43}$$

that is a function of the following dimensionless parameters, $\bar{s}$, $\bar{\tau}_2$, and $\bar{\tau}_3$. Note that the dimensionless time is given as $\bar{t} = t/\tau_1$.

A state space model of system in Figure 3.1 is given by

$$\begin{bmatrix} \ddot{x} \\ \dot{x} \end{bmatrix} = \begin{bmatrix} -\frac{c}{m} & -\frac{k}{m} \\ 1 & 0 \end{bmatrix} \begin{bmatrix} \dot{x} \\ x \end{bmatrix} + \begin{bmatrix} \frac{1}{m} \\ 0 \end{bmatrix} f \tag{3.44}$$

$$y = \begin{bmatrix} 0 & 1 \end{bmatrix} \begin{bmatrix} \dot{x} \\ x \end{bmatrix} \tag{3.45}$$

From the dimensional set of Table 3.6, we can obtain the dimensionless signals as

$$\bar{t} = t\sqrt{\frac{k}{m}}$$

$$\dot{\bar{x}} = \frac{\dot{x}}{x_0}\sqrt{\frac{m}{k}}$$

Table 3.6 Transfer function dimensional set

	t	$\dot{x}$	x	c	f	x_0	k	m
M	0	0	0	1	1	0	1	1
L	0	1	1	0	1	1	0	0
T	1	−1	0	−1	−2	0	−2	0
$\bar{t}$	1	0	0	0	0	0	1/2	−1/2
$\dot{\bar{x}}$	0	1	0	0	0	−1	−1/2	1/2
$\bar{x}$	0	0	1	0	0	−1	0	0
$\bar{c}$	0	0	0	1	0	0	−1/2	−1/2
$\bar{f}$	0	0	0	0	1	−1	−1	0

$$\bar{x} = \frac{x}{x_0}$$

$$\bar{c} = c\sqrt{\frac{1}{mk}}$$

$$\bar{f} = f\frac{1}{x_0 k}$$

The resulting dimensionless matrices are

$$N_x = \begin{bmatrix} x_0\sqrt{\dfrac{k}{m}} & 0 \\ 0 & x_0 \end{bmatrix}$$

$$N_u = \begin{bmatrix} x_0 k \end{bmatrix}$$

$$N_y = \begin{bmatrix} x_0 \end{bmatrix}$$

$$N_t = \begin{bmatrix} \sqrt{\dfrac{m}{k}} \end{bmatrix}$$

That finally yields the dimensionless representation of the state space inverted pendulum

$$\begin{bmatrix} \ddot{\bar{x}} \\ \dot{\bar{x}} \end{bmatrix} = \begin{bmatrix} -\bar{c} & -1 \\ 1 & 0 \end{bmatrix} \begin{bmatrix} \dot{\bar{x}} \\ \bar{x} \end{bmatrix} + \begin{bmatrix} 1 \\ 0 \end{bmatrix} \bar{u} \tag{3.46}$$

$$\bar{y} = \begin{bmatrix} 0 & 1 \end{bmatrix} \begin{bmatrix} \dot{\bar{x}} \\ \bar{x} \end{bmatrix} \tag{3.47}$$

Summarizing, the transfer function dimensionless representation relates the dimensional input and dimensional output divided by the transfer function gain as can be seen in (3.41) and (3.42). Furthermore, the time scale is given by the dominant time constant as $\bar{t} = t/\tau_1$.

In the state space dimensionless representation all signals are dimensionless and the time scale depends on the physical parameters of the problem at hand. Note, however, that just one dimensionless parameter is necessary for state space model $\bar{c}$ while two parameters, $\bar{\tau}_2$ and $\bar{\tau}_3$, are required for the transfer function approach.

3.5 Discrete time models dimensionless representation

Heretofore, we have considered only dimensionless representations of continuous time models. However, discrete time models are prevalent in control system applications. Therefore, in this section we derive the dimensionless representation of discrete time dynamical systems in the transfer function framework and in the state space representation.

3.5.1 Discrete time transfer function dimensionless representation

The dimensions of the discrete time transfer function parameters are given by the following lemma.

Lemma 3.2. *Consider a generic discrete time transfer function $G(z)$ in polynomial form*

$$G(z) = k\frac{\sum_{j=0}^{m}(b_j z^{-j})z^{-q}}{\left(1+\sum_{i=1}^{n}(a_i z^{-i})\right)} \tag{3.48}$$

with q integer and transfer function parameters k, b_j, and a_i. The relative system order is $n+q-m \geq 0$. The dimensions of the transfer function parameters are $[k] = K$, $[a_j] = 1$, and $[b_i] = 1$, where K is the gain dimension and 1 is a dimensionless quantity.

Proof. Consider the difference equation of transfer function (3.48)

$$\begin{aligned} y(k) &= -a_1 y(k-1) - \cdots - a_n y(k-n) \\ &\quad + k(b_0 u(k-q) + \cdots + b_m u(k-m-q)) \end{aligned}$$

by dimensional homogeneity considerations [6], the dimensions of $[a_i y(k-i)]$ must be equal to $[y(k)]$, then $[a_i y(k-i)] = Y$. It then follows that $[a_i] = 1$. In a similar way $[kb_j u(k-j-q)] = Y$, being Y the output dimension. Note also that $[u(k-j-q)] = U$, being U the input dimension, then $[kb_j] = YU^{-1} = K$. As a result, a possible dimensional decomposition of the parameter kb_j into a gain k and a parameter b_j is $[k] = K$ and $[b_j] = 1$. □

The number of transfer function variables is $N_v = n+m+3$, and the number of variables with independent dimensions is $N_d = 1$. As a result there are $n = m+n+2$ dimensionless numbers, which are

- The dimensionless number $\Pi_{G(z)} = \frac{G(z)}{k}$ is the dimensionless transfer function.
- The dimensionless number $\Pi_q = q$.

- The dimensionless numbers $\Pi_{b_j} = b_j \ j \in [1, \ldots, m]$.
- The dimensionless numbers $\Pi_{a_i} = a_i \ i \in [1, \ldots, n]$.

The dimensionless discrete transfer function is given by

$$\frac{G(z)}{k} = \frac{\sum_{j=0}^{m} (b_j z^{-j} z^{-q})}{\left(1 + \sum_{i=1}^{n} (a_i z^{-i})\right)} \tag{3.49}$$

As a result, a discrete transfer function is made dimensionless by just dividing by k.

3.5.2 Discrete time state space model dimensionless representation

Consider the following discrete time state space model.

$$x_{k+1} = A_d x_k + B_d u_k \tag{3.50}$$

$$y_k = C_d x_k + D_d u_k \tag{3.51}$$

In this case the dimensionless discrete state space model is given in a similar manner as in the continuous time case. However, in the discrete time case, despite its name, there is no time, just a sequence of samples. As a result, there is no time scaling factor as the dimensions of x_{k+1} are equal to the dimensions of x_k because there is no time derivative, just a sequence index increase. The dimensionless discrete time state space model is given by

$$\bar{A}_d = N_x^{-1} A_d N_x \tag{3.52}$$

$$\bar{B}_d = N_x^{-1} B_d N_u \tag{3.53}$$

$$\bar{C}_d = N_y^{-1} C_d N_x \tag{3.54}$$

$$\bar{D}_d = N_y^{-1} D_d N_u \tag{3.55}$$

with

$$x_k = N_x \bar{x}_k \tag{3.56}$$

$$u_k = N_u \bar{u}_k \tag{3.57}$$

$$y_k = N_y \bar{y}_k \tag{3.58}$$

In the discrete time case, the dimensionless transformation is equal to a coordinate transformation plus a signal scaling, but there is no time scaling.

Example 3.4. Consider the dynamics of a rabbit–wolf population, given by the following state space model:

$$\begin{bmatrix} r_{k+1} \\ w_{k+1} \end{bmatrix} = \begin{bmatrix} 1 + b_r & -d_r \\ b_w & 1 - d_w \end{bmatrix} \begin{bmatrix} r_k \\ w_k \end{bmatrix} + \begin{bmatrix} 0 \\ 1 \end{bmatrix} u \tag{3.59}$$

$$y = \begin{bmatrix} 0 & 1 \end{bmatrix} \begin{bmatrix} r_k \\ w_k \end{bmatrix} \tag{3.60}$$

where r_k is the rabbit population at instant k, w_k is the wolf population at instant k, b_r is the rabbit birth rate, b_w is the wolf birth rate, d_r is the rabbit death rate, and d_w is the wolf death rate. The control action u is the number of wolves set free at each instant k. w_0 is the initial number of wolves.

Table 3.7 Transfer function dimensional set

	r_k	w_k	u	b_r	d_w	d_r	w_0	b_w
Q_r	1	0	0	0	0	1	0	−1
Q_w	0	1	1	0	0	−1	1	1
$\bar{r}_k$	1	0	0	0	0	0	−1	1
$\bar{w}_k$	0	1	0	0	0	0	−1	0
$\bar{u}$	0	0	1	0	0	0	−1	0
$\bar{b}_r$	0	0	0	1	0	0	0	0
$\bar{d}_w$	0	0	0	0	1	0	0	0
$\bar{d}_r$	0	0	0	0	0	1	0	1

First, from the dimensional set of Table 3.7, we can obtain the dimensionless signals as

$$\begin{aligned}
\bar{r}_k &= r_k \frac{b_w}{w_0} \\
\bar{w}_k &= w_k \frac{1}{w_0} \\
\bar{u} &= u \frac{1}{w_0} \\
\bar{b}_r &= b_r \\
\bar{d}_w &= d_w \\
\bar{d}_r &= d_r b_w
\end{aligned}$$

The resulting dimensionless matrices are

$$\begin{aligned}
N_x &= \begin{bmatrix} x_0\sqrt{\dfrac{w_0}{b_w}} & 0 \\ 0 & w_0 \end{bmatrix} \\
N_u &= \begin{bmatrix} w_0 \end{bmatrix} \\
N_y &= \begin{bmatrix} w_0 \end{bmatrix}
\end{aligned}$$

That finally yields the dimensionless representation

$$\begin{bmatrix} \bar{r}_{k+1} \\ \bar{w}_{k+1} \end{bmatrix} = \begin{bmatrix} 1+b_r & -\bar{d}_r \\ 1 & 1-d_w \end{bmatrix} \begin{bmatrix} \bar{r}_k \\ \bar{w}_k \end{bmatrix} + \begin{bmatrix} 0 \\ 1 \end{bmatrix} \bar{u} \tag{3.61}$$

$$\bar{y} = \begin{bmatrix} 0 & 1 \end{bmatrix} \begin{bmatrix} \bar{r}_k \\ \bar{w}_k \end{bmatrix} \tag{3.62}$$

3.6 Exercises

1. Consider the following transfer functions commonly used in process control:

$$G_1(s) = \frac{k}{(\tau_1 s + 1)} e^{hs}$$

$$G_2(s) = \frac{k}{(\tau_1 s + 1)(\tau_2 s + 1)}$$

$$G_3(s) = \frac{k}{(\tau_1 s + 1)(\tau_2 s + 1)} e^{hs}$$

$$G_4(s) = \frac{k(\tau 3s + 1)}{(\tau_1 s + 1)(\tau_2 s + 1)}$$

$$G_5(s) = \frac{k(\tau 3s + 1)}{(\tau_1 s + 1)(\tau_2 s + 1)} e^{hs}$$

   a) Calculate the number of dimensionless number for each transfer function.
   b) Considering the number of dimensionless numbers a measure of complexity, classify the transfer functions in terms of complexity.
   c) Obtain the dimensionless representation of the transfer functions.
2. The state space model of an active band-pass filter is

$$\left[\frac{dx}{dt}\right] = \begin{bmatrix} \frac{-1}{r_1 c_1} & 0 \\ \frac{1}{r_1 c_2} & \frac{-1}{r_2 c_2} \end{bmatrix} [x] + \begin{bmatrix} \frac{1}{r_1 c_1} \\ \frac{-1}{r_1 c_2} \end{bmatrix} u \tag{3.63}$$

$$y = \begin{bmatrix} 0 & 1 \end{bmatrix} [x] \tag{3.64}$$

   with r_1, r_2 electrical resistances, c_1, c_2 capacitances, and the state x and u voltages.
   a) Obtain the dimensionless state space form.
   b) Obtain the transfer function as $G(s) = C(sI - A)^{-1}B$.
   c) Calculate the dimensionless transfer function representation.
   d) Simulate the output for a step input for the dimensionless state space model and for the dimensionless transfer function and compare the time scales of each dimensionless model.

References

[1] P. Balaguer, A. Ibeas, C. Pedret, and S. Alcántara. 'Controller parameters dependence on model information through dimensional analysis'. In *Decision and Control, 2009 held jointly with the 2009 28th Chinese Control Conference. CDC/CCC 2009. Proceedings of the 48th IEEE Conference on*, pages 1914–1919. IEEE, 2009

[2] S.N. Brennan. *On size and control: the use of dimensional analysis in controller design*. PhD thesis, Citeseer, 2002

[3] K. Ogata. *Modern Control Engineering*. Prentice Hall PTR, 2001

[4] H.J. Palanthandalam-Madapusi, D.S. Bernstein, and R. Venugopal. Dimensional analysis of matrices state-space models and dimensionless units [lecture notes]. *Control Systems Magazine, IEEE*, 27(6):100–109, 2007

[5] S. Skogestad and I. Postlethwaite. *Multivariable Feedback Control. Analysis and Design*. Wiley, 1996

[6] T. Szirtes and P. Rózsa. *Applied Dimensional Analysis and Modeling*. Butterworth-Heinemann, 2007

[7] B. West. 'The tyranny of many dimensionless constants: a constraint on knowability'. *Uncertainty and Surprise in Complex Systems*, Springer, pages 87–92, 2005

Chapter 4
Dynamical systems: dimensional similarity

4.1 Introduction

The dimensionless representation of dynamical systems presented in the previous chapter is the fingerprint for establishing dimensional similarity among dynamical systems. The concept of dimensional similarity is of great utility in model experimentation as it permits to extrapolate the model experimental results to the prototype. Furthermore, even the theoretical results obtained in dimensionless systems can be easily extrapolated to dimensionally similar models [2].

In this chapter we present the conditions for establishing dimensional similarity between systems represented in transfer function form and in state space form. It is easier to establish dimensional similarity between transfer functions than between state space systems representation. The transfer function framework only requires two scaling factors between model and prototype, the gain scaling factor and the time scaling factor. On the contrary, in the state space framework, there are as many scaling factors as physical variables involved in the matrices representation, thus the scaling factors are problem dependent.

The second half of the chapter is devoted to dimensional similarity between discrete time systems, considering also the transfer function and the state space representation. With discrete time systems it is necessary to stress between purely discrete time systems and the discrete time systems obtained from continuous time systems by discretization. For purely discrete time systems there is no time, and dimensional similarity is reduced to model equality up to a gain factor. For sampled data systems, the time concept is recovered and for preserving continuous time dimensional similarity the sampling times used must also be scaled properly.

4.2 Continuous time dynamical systems similarity

4.2.1 Transfer function dimensional similarity

The definition of transfer function dimensional similarity follows directly from its dimensionless representation presented in Chapter 3.

Definition 4.1. Two transfer functions, the prototype $G_p(s_p)$ and the model $G_m(s_m)$, are similar if the following dimensionless numbers are equal:

$$\left(\frac{\tau_{pi}}{\tau_{p1}}\right)_p = \left(\frac{\tau_{pi}}{\tau_{p1}}\right)_m, \quad i = 2, \ldots, n, \tag{4.1}$$

$$\left(\frac{\tau_{cj}}{\tau_{p1}}\right)_p = \left(\frac{\tau_{cj}}{\tau_{p1}}\right)_m, \quad j = 1, \ldots, m \tag{4.2}$$

$$\left(\frac{h}{\tau_{p1}}\right)_p = \left(\frac{h}{\tau_{p1}}\right)_m \tag{4.3}$$

$$\left(\tau_{p1}s\right)_p = \left(\tau_{p1}s\right)_m \tag{4.4}$$

$$(l)_p = (l)_m \tag{4.5}$$

with τ_{p1} the dominant time constant, $(\,.\,)_p$ stands for the prototype and $(\,.\,)_m$ stands for the model.

The same prototype and model transfer function structure is a necessary but not sufficient condition for transfer function similarity, because dimensionless numbers must be equal in construction and in magnitude. As a result, in the rest of the chapter we assume that $(l)_p = (l)_m$, that is the numbers of integrators in prototype and model are equal.

Recall that the scaling factor of a particular physical variable is defined as the quotient of the prototype and the model physical magnitude. Given a prototype and a model transfer function we have, for instance, the gain scaling factor $S_k := (k)_p/(k)_m$, associated with the transfer function gain, or, the time delay scaling factor $S_h := (h)_p/(h)_m$, associated with the transfer function time delay. For all the remaining parameters we could consider their corresponding scaling factor. However, we show next that when two transfer functions are similar, there are only two scaling factors, the time scaling factor S_t and the gain scaling factor S_k. First we need the following lemma.

Lemma 4.1. *The scaling factor associated with any transfer function parameter with time dimension is equal to* $S_t := (\tau_{p1})_p/(\tau_{p1})_m$*, where* $(\tau_{p1})_p$ *and* $(\tau_{p1})_m$ *are the prototype and model dominant time constant, respectively.*

Proof. First consider the time delay scaling factor S_h. From (4.3) it follows that

$$S_h := \frac{(h)_p}{(h)_m} = \frac{(\tau_{p1})_p}{(\tau_{p1})_m} \tag{4.6}$$

The same reasoning can be applied to all the time constants presented in (4.1) and (4.2). Finally the dimensionless number involving frequency is given by $\tau_{p1}s$. From (4.4) it can be derived that

$$S_s := \frac{(s)_p}{(s)_m} = \frac{(\tau_{p1})_m}{(\tau_{p1})_p} \tag{4.7}$$

or in the time domain

$$S_t := \frac{t_p}{t_m} = \frac{(\tau_{p1})_p}{(\tau_{p1})_m} \tag{4.8}$$

As a result the time scaling factor S_t is the same and equal to $(\tau_{p1})_p/(\tau_{p1})_m$ for every transfer function parameter with time dimension. □

Recall now that if all dimensionless numbers related to governing variables are equal, then the dimensionless numbers related to the governed variables of prototype and model are also equal, that is

$$\left(\frac{G(s)}{k_p\tau_{p1}^l}\right)_p = \left(\frac{G(s)}{k_p\tau_{p1}^l}\right)_m \tag{4.9}$$

which can be rewritten as

$$(G(s))_p = \frac{\left(k_p\tau_{p1}^l\right)_p}{\left(k_p\tau_{p1}^l\right)_m}(G(s))_m \tag{4.10}$$

and by definition of the gain and time scaling factors it yields

$$(G(s))_p = S_k S_t^l\,(G(s))_m \tag{4.11}$$

that provides the relation between dimensionally similar transfer functions in terms of gain and time scaling factors. Furthermore, the transformations between the model transfer function $G_m(s_m) := (G(s))_m$, the prototype transfer function $G_p(s_p) := (G(s))_p$ and the dimensionless transfer function $\bar{G}(\bar{s})$ are summarized in Figure 4.1. For instance, the prototype transfer function is obtained from the model transfer function by just scaling with $S_k S_t^l$ and the change of variable $s_m \to S_t s_p$.

Example 4.1. As an example, consider two first-order plus dead-time (FOPDT) transfer functions:

$$G_p(s_p) = \frac{k_p}{(\tau_p + 1)}e^{h_p s_p} \tag{4.12}$$

$$G_m(s_m) = \frac{k_m}{(\tau_m + 1)}e^{h_m s_m} \tag{4.13}$$

where s_p and s_m are the Laplace variables of prototype and model, respectively. $G_p(s_p)$ and $G_m(s_m)$ are dimensionally similar if $h_p/\tau_p = h_m/\tau_m$ and $\tau_p s_p = \tau_m s_m$.

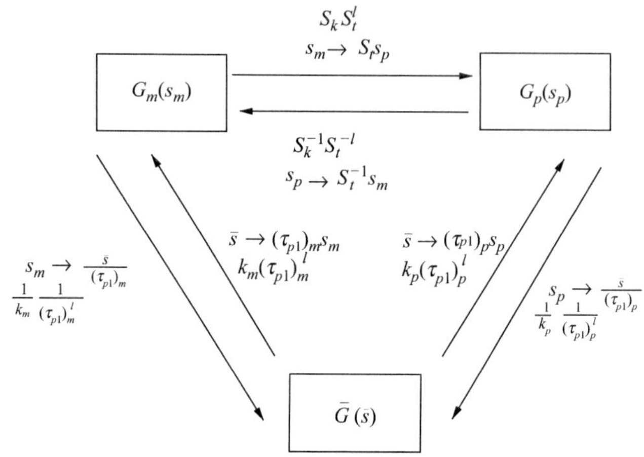

Figure 4.1 Transformations between model $G_m(s_m)$, prototype $G_p(s_p)$ and dimensionless transfer function $\bar{G}(\bar{s})$. The dimensionless form is obtained by gain scaling and time scaling. The relation between model and prototype is defined by means of two scaling factors, the gain scaling factor S_k and the time scaling factor S_t

In what follows $\bar{s}$ refers to either $\tau_p s_p$ or $\tau_m s_m$. It follows that $G_p(\bar{s})/k_p = G_m(\bar{s})/k_m$, or equivalently

$$G_p(\bar{s}) = S_k G_m(\bar{s}) \tag{4.14}$$

with $S_k = (k)_p/(k)_m$, the gain scaling factor.

4.2.2 State space dimensional similarity

The dimensionless representation of dynamical systems in state space form is given as

$$\dot{\bar{x}}(\bar{t}) = \bar{A}\bar{x}(\bar{t}) + \bar{B}\bar{u}(\bar{t}) \tag{4.15}$$

$$\bar{y}(\bar{t}) = \bar{C}\bar{x}(\bar{t}) + \bar{D}\bar{u}(\bar{t}) \tag{4.16}$$

where the dimensionless matrices $\bar{A}$, $\bar{B}$, $\bar{C}$, and $\bar{D}$ are functions of dimensionless numbers related to governing variables. In this way we can write

$$\bar{A} = \bar{A}(\Pi_{A1}, \ldots, \Pi_{An})$$
$$\bar{B} = \bar{B}(\Pi_{B1}, \ldots, \Pi_{Bm})$$
$$\bar{C} = \bar{C}(\Pi_{C1}, \ldots, \Pi_{Cp})$$
$$\bar{D} = \bar{D}(\Pi_{D1}, \ldots, \Pi_{Dq})$$

where there are $n+m+p+q$ dimensionless numbers related to governing variables. In this case, there are as many scaling factors as governing variables, and they are problem dependent. Dimensional similarity is established if

$$(\bar{A})_p = (\bar{A})_m \tag{4.17}$$

$$(\bar{B})_p = (\bar{B})_m \tag{4.18}$$

$$(\bar{C})_p = (\bar{C})_m \tag{4.19}$$

$$(\bar{D})_p = (\bar{D})_m \tag{4.20}$$

Example 4.2. Consider again the dimensionless representation of the state space inverted pendulum

$$\begin{bmatrix} \ddot{\bar{\theta}} \\ \dot{\bar{\theta}} \end{bmatrix} = \begin{bmatrix} -\bar{c} & \bar{g}\bar{l}-1 \\ 1 & 0 \end{bmatrix} \begin{bmatrix} \dot{\bar{\theta}} \\ \bar{\theta} \end{bmatrix} + \begin{bmatrix} 1 \\ 0 \end{bmatrix} \bar{u} \tag{4.21}$$

$$\bar{y} = \begin{bmatrix} 0 & 1 \end{bmatrix} \begin{bmatrix} \dot{\bar{\theta}} \\ \bar{\theta} \end{bmatrix} \tag{4.22}$$

with

$$\bar{g} = g\frac{\sqrt{im}}{k}$$

$$\bar{l} = l\sqrt{\frac{m}{i}}$$

$$\bar{c} = c\sqrt{\frac{1}{ik}}$$

We have one scaling factor per governing variable, that is a gravity acceleration scaling factor S_g, a length scaling factor S_l, a damping scaling factor S_d, a mass scaling factor S_m, an inertia scaling factor S_i, and a stiffness scaling factor S_k.

In this case, for dimensional similarity it must follow that

$$\left(c\sqrt{\frac{1}{ik}}\right)_p = \left(c\sqrt{\frac{1}{ik}}\right)_m \tag{4.23}$$

$$\left(g\frac{\sqrt{im}}{k}l\sqrt{\frac{m}{i}}\right)_p = \left(g\frac{\sqrt{im}}{k}l\sqrt{\frac{m}{i}}\right)_m \tag{4.24}$$

that is equivalent to the following model laws:

$$S_c = \sqrt{S_i S_k} \tag{4.25}$$

$$S_l = \frac{S_k}{S_m S_g} \tag{4.26}$$

Summing up, to establish dimensional similarity between prototype and model in the state space framework is more complicated than the transfer function case because prototype and models are related by more than two scaling factors.

4.3 Discrete time dynamical system similarity

In this section we establish the similarity properties of discrete transfer functions and sampled data transfer functions discretized from similar continuous time systems.

4.3.1 Discrete time transfer function similarity

Definition 4.2. Two discrete time transfer functions, the prototype $G_p(z)$ and the model $G_m(z)$, are dimensionally similar if the following dimensionless numbers are equal:

$$\left(\frac{G(z)}{k}\right)_p = \left(\frac{G(z)}{k}\right)_m \tag{4.27}$$

$$(q)_p = (q)_m \tag{4.28}$$

$$(b_j)_p = (b_j)_m, \quad j = 1, \ldots, m \tag{4.29}$$

$$(a_i)_p = (a_i)_m, \quad i = 1, \ldots, n \tag{4.30}$$

In this case dimensional similarity requires equality between the prototype and model transfer function, except for a gain constant. As a result there is just one scaling factor, the gain scaling factor $S_k := (k)_p/(k)_m$. It then follows that under dimensional similarity model and prototype are related by

$$G_p(z) = S_k G_m(z) \tag{4.31}$$

4.3.2 Sampled-data transfer function similarity

A sampled-data transfer function $G(z_{SD})$ is a discrete time transfer function obtained by discretization of a continuous time transfer function $G(s)$. Whereas the discrete time transfer function variable z is related to the shift operator with no time relation whatsoever, the sampled data transfer function variable z_{SD} is related with the continuous time variable by means of $z = e^{T_s s}$, where T_s is the sampling time [1].

Definition 4.3. Two sampled data transfer functions $G_p(z_{SD})$ and $G_m(z_{SD})$, discretized with sampling time $(T_s)_p$ and $(T_s)_m$, are dimensionally similar if (i) $G_p(z_{SD})$ and $G_m(z_{SD})$ are discrete time dimensionally similar and (ii) $(e^{sT_s})_p = (e^{sT_s})_m$.

Note that sampled data similarity implies discrete time similarity, but the converse is not true. The requirements in the discretization process in order to yield similar sampled-data transfer functions are derived next.

Theorem 4.1. *Consider two continuous time transfer functions $G_p(s)$ and $G_m(s)$ with transfer function time scaling factor S_t, discretized using sampling times $(T_s)_p$ and $(T_s)_m$, with $(T_s)_p = S_t(T_s)_m$, yielding the sampled data transfer functions $G_p(z_{SD})$ and $G_m(z_{SD})$, respectively, then if $G_p(s)$ and $G_m(s)$ are continuous time dimensionally similar, then $G_p(z_{SD})$ and $G_m(z_{SD})$ are sampled-data dimensionally similar.*

Proof. By Definition 4.2, two discrete time transfer functions are dimensionally similar if they have the same dimensionless numbers, that is $(q)_p = (q)_m$; $(b_j)_p = (b_j)_m$, $j = 0, \ldots, m$; and $(a_i)_p = (a_i)_m$, $i = 1, \ldots, n$. The parameters b_j and a_i are obtained by the discretization process Φ, which can be, for instance, the zero order hold method or Tustin method. The discretization depends on the continuous time transfer function parameters and on the sampling time T_s, thus

$$a_i = \Phi(\tau_{p1}, \ldots, \tau_{pn}, \tau_{c1}, \ldots, \tau_{cm}, h, l, T_s) \tag{4.32}$$

which has the following dimensionless form:

$$a_i = \bar{\Phi}\left(\frac{\tau_{p2}}{\tau_{p1}}, \ldots, \frac{\tau_{pn}}{\tau_{p1}}, \frac{\tau_{c1}}{\tau_{p1}}, \ldots, \frac{\tau_{cm}}{\tau_{p1}}, \frac{h}{\tau_{p1}}, \frac{l}{\tau_{p1}}, \frac{T_s}{\tau_{p1}}\right) \tag{4.33}$$

If $G_p(s)$ and $G_m(s)$ are continuous time dimensionally similar and the sampling time fulfils the condition $(T_s)_p = S_t(T_s)_m$, then all the dimensionless numbers in (4.33) are equal for the prototype and the model, yielding $(a_i)_m = (a_i)_p$. The equality $(b_j)_m = (b_j)_p$ is established by similar reasoning. The discrete time delay is related with the continuous time delay and the sampling time by $q = h/T_s$. In order to have $(q)_p = (q)_m$ it must follow that $\frac{h_p}{h_m} = \frac{(T_s)_p}{(T_s)_m}$, which is true by continuous time similarity. Finally, for achieving sampled-data similarity it must also hold the equality in the dimensionless number z_{SD}, that is $(z_{SD})_m = (z_{SD})_p$. The equality follows by recalling the relation existing between the s transform and the z transform, given by $z = e^{sT_s}$, and the assumption of continuous time similarity, which implies that if $e^{s(T_s)_p} = e^{s(T_s)_m}$ then $(z_{SD})_p = (z_{SD})_m$. □

Example 4.3. Consider the two continuous time transfer functions presented in the first column of Table 4.1. They are dimensionally similar with similarity scale factors $S_k = 10$ and $S_t = 0.8$. Both transfer functions are discretized by ZOH with sampling times $(T_s)_m = 0.1$ and $(T_s)_p = 0.08$, respectively. It can be seen in the second column of Table 4.1 that they are sampled-data dimensionally similar.

4.3.3 Discrete state space similarity

The dimensionless representation of a discrete state space model is

$$\bar{x}_{k+1} = \bar{A}_d\bar{x}_k + \bar{B}_d\bar{u}_k \tag{4.34}$$

$$\bar{y}_k = \bar{C}_d\bar{x}_k + \bar{D}_d\bar{u}_k \tag{4.35}$$

Table 4.1 Example of sampled data transfer function similarity

	Similar continuous time	Similar discrete time
G_m	$\frac{0.3s^2+30.01s+1}{s^3+3s^2+3s+1}$	$\frac{0.1617z^2-0.06392z-0.09693}{z^3-2.715z^2+2.456z-0.7408}$
G_p	$10\frac{0.192s^2+24.01s+1}{0.512s^3+1.92s^2+2.4s+1}$	$10\frac{0.1617z^2-0.06392z-0.09693}{z^3-2.715z^2+2.456z-0.7408}$

In the same way as in the continuous time case, the dimensionless matrices $\bar{A}_d$, $\bar{B}_d$, $\bar{C}_d$, and $\bar{D}_d$ are functions of dimensionless numbers related to governing variables, thus

$$\begin{aligned}
\bar{A}_d &= \bar{A}_d(\Pi_{A_d1},\ldots,\Pi_{A_dn})\\
\bar{B}_d &= \bar{B}_d(\Pi_{B_d1},\ldots,\Pi_{B_dm})\\
\bar{C}_d &= \bar{C}_d(\Pi_{C_d1},\ldots,\Pi_{C_dp})\\
\bar{D}_d &= \bar{D}_d(\Pi_{D_d1},\ldots,\Pi_{D_dq})
\end{aligned}$$

where there are $n+m+p+q$ dimensionless numbers related with governing variables. Dimensional similarity is established if the dimensionless matrices of model and prototype are equal, that is

$$(\bar{A}_d)_p = (\bar{A}_d)_m \tag{4.36}$$

$$(\bar{B}_d)_p = (\bar{B}_d)_m \tag{4.37}$$

$$(\bar{C}_d)_p = (\bar{C}_d)_m \tag{4.38}$$

$$(\bar{D}_d)_p = (\bar{D}_d)_m \tag{4.39}$$

In this case, there are as many scaling factors as governing variables, and they are also problem dependent.

4.4 Exercises

1. Given the following prototype and model transfer functions:

$$G_m(s) = \frac{2(2s-1)}{(5s+1)(3s+1)(s+1)}e^{1.75s} \tag{4.40}$$

$$G_p(s) = \frac{7(1.5s-1)}{(3.75s+1)(2.25s+1)(0.75s+1)}e^{1.5s} \tag{4.41}$$

a) Determine if they are dimensionally similar.
b) Modify the model time delay in order to achieve complete similarity.

c) Find the relationship between sampling times that preserve dimensional similarity.
d) Obtain the zero order hold discretization of continuous time transfer functions with previous sampling times ratio and check the equality of discrete time transfer functions up to a gain factor.

2. The state space representation of an inverted pendulum is

$$\begin{bmatrix} \ddot{\theta} \\ \dot{\theta} \end{bmatrix} = \begin{bmatrix} -\dfrac{c}{i} & \dfrac{mgl - k}{i} \\ 1 & 0 \end{bmatrix} \begin{bmatrix} \dot{\theta} \\ \theta \end{bmatrix} + \begin{bmatrix} \dfrac{1}{i} \\ 0 \end{bmatrix} u \tag{4.42}$$

$$y = \begin{bmatrix} 0 & 1 \end{bmatrix} \begin{bmatrix} \dot{\theta} \\ \theta \end{bmatrix} \tag{4.43}$$

with θ the angular position, $\dot{\theta}$ the angular velocity, u the input torque, g the gravity acceleration, l the pendulum length, k the angular spring stiffness, c the friction constant, m the pendulum mass, and i the pendulum inertia.

a) Obtain a dimensionless description and derive the model laws.
b) Repeat the exercise in case the inertia is given by $i = ml^2$.

References

[1] G.F. Franklin, J.D. Powell, and M.L. Workman. *Digital Control of Dynamic Systems*. Prentice Hall, 1997

[2] M. Zlokarnik. *Scale-up in Chemical Engineering*. Wiley-VCH, 2002

Chapter 5

Dimensionless systems identification and model order reduction

with J. Romero, V. Alfaro and O. Arrieta

5.1 Introduction

The identification of models from experimental data is the first step in model-based control design. Furthermore, when the model is of high order, it is normal to reduce the model order for a simpler controller design. Low order identified models are the fundamental elements for the design of practical control systems.

Although there are well-established techniques of general application for model identification and model order reduction, such as minimum squares and balanced model reduction, they are based on numerical calculations and give no parameter value insight. Particularized algorithms for system identification and model reduction can provide further understanding on the problem at hand based on the relative value among parameters.

The development of system identification and model reduction algorithms faces two difficulties. First the relations between transfer function parameters and time domain data are complex, and it is not straightforward to obtain model parameters from data, especially when the parameters number increases. For instance, approaches for the identification of integral second order systems showing inverse response have been based on the solution of a set of nonlinear equations [6] or the numerical evaluation of an integral [11]. The second difficulty faced is to evaluate what parameters can be dismissed without compromising the model accuracy. Normally, it is not correct to dismiss large or small parameters because, due to the Buckingham pi theorem, what influences the model behavior is not their absolute value but the relative value among parameters. In the following section we show how dimensional analysis helps to circumvent these problems.

5.2 General procedure

The dimensionless representation of dynamical systems reduces the number of parameters of the dynamical system. In this way we may gain some simplicity on the system representation. The parameters reduction by the dimensionless transfer function representation is limited to two parameters. Despite the limit on the parameters reduction capabilities, the dimensionless representation may enable the graphical representation in two or three dimensions of expressions that, otherwise, would not be possible to represent graphically. This feature may help in developing new identification procedures as it is shown in the identification of system with inverse response.

There is, however, another advantage in representing a dynamical system by means of dimensionless numbers. The dimensionless numbers obtained may provide further physical insight than the dimensional counterpart. In case that certain dimensionless numbers, by their construction and their physical meaning, are likely to be large and/or small, simpler relations may be established.

The general procedure to apply the dimensional analysis for system identification and model order reduction is the following:

1. The starting point is a dimensionless representation of the system to be identified or reduced.
2. Analyse the dimensionless parameters obtained, looking for large and small dimensionless numbers.
3. Discard the too large or small dimensionless parameters.
4. Formulate useful relations for identification or model reduction using the remaining dimensionless numbers.

The general procedure is applied to two particular problems to show an example of application. Both problems are not easily solved by other means. The first example of application is the identification of second order system with inverse response plus time delay. The second application example is the model order reduction for decentralized control.

5.3 Example 1: Second order inverse response model identification

5.3.1 Problem statement

Inverse response systems are present in several industrial and chemical processes, such as boilers, chemical reactors, etc. [7]. These processes present serious challenges for control design because of their nature of competing dynamics and they impose fundamental limitations to the performance and robustness attainable levels [10]. Accurate and easy-to-perform identification techniques are then necessary in order to control and establish fundamental limitations of these processes.

The objective is to derive an identification algorithm for inverse response systems with two poles and a time delay (see (5.1)), based on the process reaction curve. The assumed structure of the plant to be identified is

$$G(s) = \frac{k(-\tau_3 s + 1)}{(\tau_1 s + 1)(\tau_2 s + 1)} e^{-hs} \tag{5.1}$$

Although in the literature it is possible to find identification algorithms for systems described by (5.1), they are complex because they are based on complex experimental data in time and frequency domains [3]. Furthermore, the case of inverse response of integrating processes has also been investigated. The work of Reference 6 provides a systematic approach to obtain all the three parameters for integrating processes with dead time and inverse response. It is accomplished by numerically solving a set of three nonlinear equations. Another approach for integrating processes is proposed in Reference 11 in which the zero time constant is identified in closed form by means of an analytic expression and the system gain and the pole time constant are calculated by means of numerical integral evaluation.

Previous identification procedures are complex due to the data required and their operations. However, simpler identification algorithms based on simple experiments are well accepted in industry because normally simple models are pursued for control purposes and the economical impact due to experimentation is minimized.

In what follows we make use of the dimensionless representation of model (5.1) to derive a very simple identification algorithm. The key point for the algorithm is to realize that certain dimensionless number a is normally small in second order inverse response systems, due to the competing dynamics of these systems.

5.3.2 *Dimensionless representation of second order inverse response model*

The dimensionless representation of transfer function (5.1) is

$$\bar{G}(\bar{s}) := \frac{G(\bar{s})}{k} = \frac{(-b\bar{s}+1)}{(\bar{s}+1)(a\bar{s}+1)}e^{-c\bar{s}} \tag{5.2}$$

where

- $\bar{s} = \tau_1 s$, which can be interpreted either as a time scaling or as a frequency scaling.
- a is the ratio between the fast and slow time constants, that is $a = \tau_2/\tau_1$. The values of a considered are $a \in [0.1, \ldots, 0.5]$.
- b is the ratio between the zero time constant and the slow time constant, that is $b = \tau_3/\tau_1$. The values of b considered are $b \in [0.1, \ldots, 4]$.
- c is the ratio between the time delay and the slow time constant, that is $c = h/\tau_1$.

The use of the dimensionless representation (5.2) is a key point enabling the identification algorithm. It allows the normalized system output to be characterized (i.e. y/k) at normalized time instants (i.e. t_a/τ_1) as a function of just three dimensionless parameters, a, b, and c. Note that the transfer function (5.1) has five parameters, so dimensional analysis permits a reduction of two parameters by the use of dimensionless numbers.

The objective is, on the basis of the sole information of, at least, three points of the step response curve, to obtain an identification algorithm that provides the values of k, h, τ_1, b, and a of transfer function (5.2), which is equivalent to identifying the values of k, h, τ_1, τ_2, and τ_3 of transfer function (5.1).

5.3.3 Identification procedure

The proposed identification algorithm provides the values of k, h τ_1, b, and a in a sequential manner by the use of, at least, three points of the system reaction curve.

Identification of the gain k

Given the total output change Δy due to an input change Δu, the system gain is determined as

$$k = \frac{\Delta y}{\Delta u} \tag{5.3}$$

Identification of the time delay h

The time delay h can be identified from the step reaction curve following graphical methods, as presented in Reference 1. In this way both the system gain k and the time delay h are identified from the reaction curve and are the basis for the identification of system dynamic behavior as discussed in the following.

Identification of the dominant time constant τ_1

The system inverse response can be understood as the interaction of slow and fast dynamics [5]. In fact the considered transfer function (5.1) can be equivalently represented as a sum of two first order systems with inverse sign gains:

$$\bar{G}(\bar{s}) = \frac{(-b\bar{s}+1)}{(\bar{s}+1)(a\bar{s}+1)}e^{-c\bar{s}} = \left(\frac{k_1}{\bar{s}+1} - \frac{k_2}{a\bar{s}+1}\right)e^{-c\bar{s}} \tag{5.4}$$

with

$$k_1 = \frac{1+b}{1-a} \tag{5.5}$$

$$k_2 = \frac{a+b}{1-a} \tag{5.6}$$

We have an inverse response system if the following condition is accomplished, which is the dimensionless counterpart of the dimensional one provided in [5]:

$$\frac{1}{a} > \frac{1+b}{a+b} > 1 \tag{5.7}$$

In physical processes showing inverse response behavior it is normally the case that the competing dynamics are different in the sense that one dynamic is much slower than the other one. In fact from expression (5.7) if both dynamics are equal (i.e. $\tau_1 = \tau_2$, or equivalently $a = 1$) the process cannot have inverse response behavior. Moreover, in case that both dynamics are similar (i.e. $a \approx 1$), that implies a very high gain of the competing dynamics k_1 and k_2 due to the denominator term $(1 - a)$ in (5.5) and (5.6). Thus, it is expected that in physical processes the cause of inverse response is mainly due to the difference of two moderate gains competing at distinct dynamics instead of two dynamically similar high gain processes.

The system step time response for absolute time values t_a greater than the system delay h, expressed by means of dimensionless constants, is given by

$$\frac{y}{k\Delta u} = 1 + \left(\frac{b-1}{1-a}\right) e^{-(t_a-h)/\tau_1} + \left(\frac{b-a}{a-1}\right) e^{-(t_a-h)/\tau_1 a} \tag{5.8}$$

which can be rewritten using the normalized output $y_{\%} = y/k\Delta u = y/\Delta y$ and the relative normalized time $t' = (t_a - h)/\tau_1$ as follows:

$$y_{\%} = 1 + \left(\frac{b-1}{1-a}\right) e^{-t'} + \left(\frac{b-a}{a-1}\right) e^{-t'/a} \tag{5.9}$$

Note that in the rest of the chapter, the time variables t' are not absolute but relative (i.e. $t' = t'_a - c$).

Assuming, as discussed above, that constant a is small (e.g. $a < 0.5$), that is both poles have well-differentiated dynamics, it implies that the term $e^{-t'/a}$ in (5.9) tends to zero faster than $e^{-t'}$. In this case it is reasonable to approximate the step response (5.9) for large enough values of time t' by

$$y_{\%} \approx \bar{y}_{\%} := 1 + \left(\frac{b-1}{1-a}\right) e^{-\bar{t}'} \tag{5.10}$$

In order to compare graphically both outputs, in Figure 5.1 we present the exact and approximate outputs for $a = 0.1$ and $a = 0.5$ (in both cases $b = -1$). As can be seen the smaller the a parameter, the better the approximation of (5.10) for smaller time values.

Before proceeding, it is important to stress the distinction between t'_x and $\bar{t}'_x$. Parameter t'_x is the normalized time for the normalized output achieving the value of $x\%$ as given by (5.9). Parameter $\bar{t}'_x$ is the normalized time for normalized output achieving the value of $x\%$ as given by approximated equation (5.10). As time increases, the value of $\bar{t}'_x$ tends to t'_x.

Now, taking logarithms from (5.10):

$$\ln(y_{\%} - 1) = \ln\left(\frac{b-1}{1-a}\right) - \bar{t}' \tag{5.11}$$

and particularizing (5.11) for two points $\{y_{x\%}, t_x\}$ and $\{y_{y\%}, t_y\}$ of the reaction curve with t_x and t_y large enough, we can identify τ_1 as

$$\tau_1 = \frac{t_y - t_x}{\ln\left(\frac{y_{x\%}-1}{y_{y\%}-1}\right)} \tag{5.12}$$

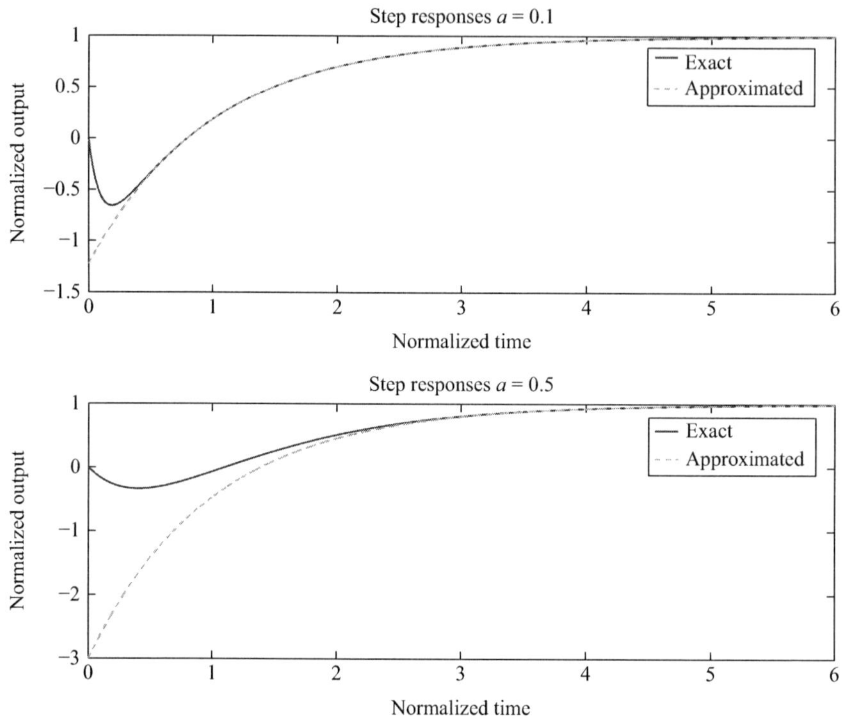

Figure 5.1 Normalized exact and approximated step responses, for a = 0.1 and a = 0.5

Identification of constant *b*

The constant b, once the dominant time constant τ_1 is known, can be identified by taking the point of minimum value of the step response (y_p, t_p). It follows that the derivative of the output at that point is zero, then

$$\frac{dy}{dt} = \left(\frac{1-b}{1-a}\right) e^{-t'_p} + \left(\frac{a-b}{a-1}\right) e^{-t'_p/a} \frac{1}{a} = 0 \tag{5.13}$$

From (5.9) and (5.13) we finally obtain

$$\bar{b} = 1 - \frac{1 - y_{p\%}}{e^{-t'_p}} \tag{5.14}$$

Identification of constant *a*

Finally constant a is identified by analyzing the variation of the relative normalized time t' from (5.9) with respect to parameters a and b. In Figure 5.2 it can be observed how the normalized time varies for output values from 10% to 90% as

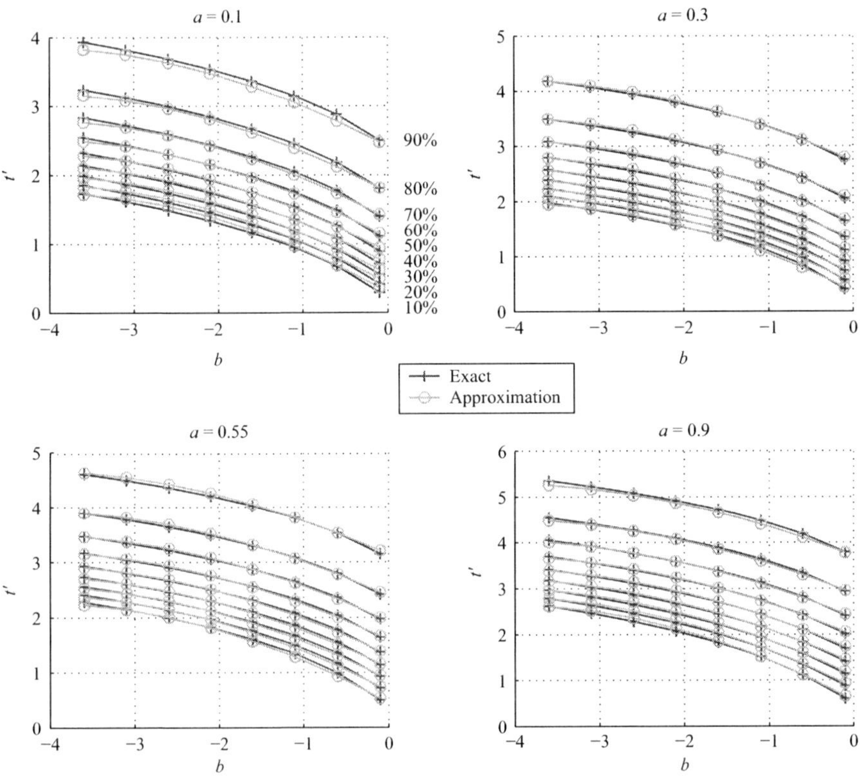

Figure 5.2 Normalized time as a function of b for a constant

a function of parameter b assuming a constant. The curves are clearly not linear; however, a quadratic form can be fairly adjusted as can be seen in Figure 5.2. On the other hand, Figure 5.3 presents the normalized time variation, also for output values between 10% and 90% as a function of parameter a assuming now a constant value of b. It can be seen that in this case the curves can be well adjusted by a linear function.

From the results of Figures 5.2 and 5.3, it is proposed to adjust the normalized time by means of the following expression:

$$t'_x = m_{1x} + m_{2x}a + (m_{3x} + m_{4x}a)b + (m_{5x} + m_{6x}a)b^2 \tag{5.15}$$

with m_{ix}, $i \in [1, 6]$ parameters adjusted by least squares. The rationale of (5.15) is that for b constant (5.15) is linear in a, while for a constant (5.15) is quadratic in b, as required from Figures 5.2 and 5.3. Moreover, it has enough degrees of freedom to capture the curvature changes as a function of a and b. In Figure 5.4 the values m_{ix}, $i \in [1, 6]$ adjusted by least square approximation can be seen. The approximation

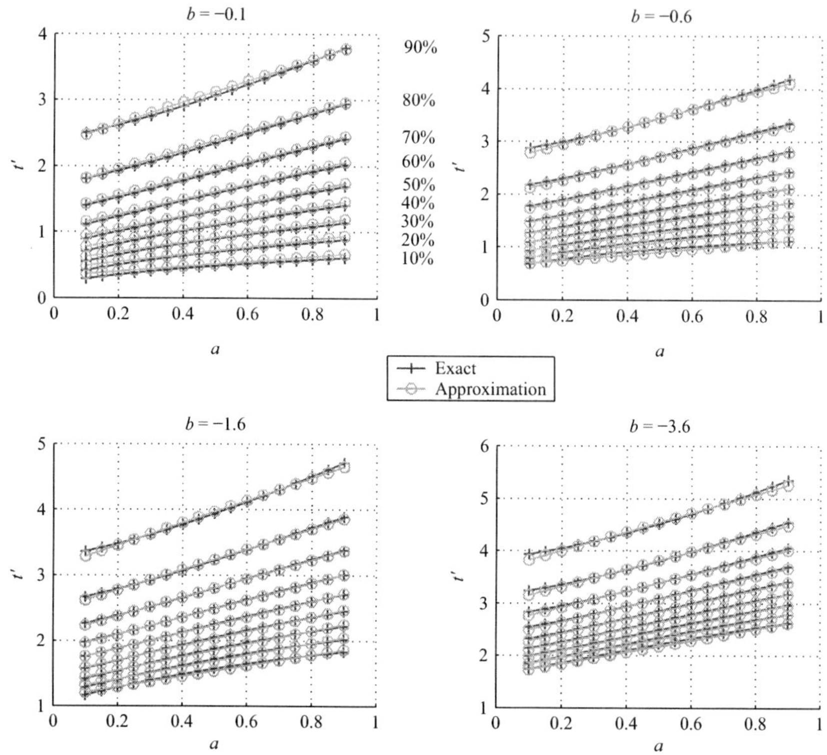

Figure 5.3 Normalized time as a function of a for b constant

is performed on the basis of the dimensionless transfer function (5.2). Thus, from function (5.15) and values b and τ_1 previously identified, a is given by

$$\bar{a} = \frac{t'_x - (m_{1x} + m_{3x}b + m_{5x}b^2)}{m_{2x} + m_{4x}b + m_{6x}b^2} \tag{5.16}$$

General identification algorithm

The general identification algorithm provides the transfer function parameters by using three or four points of the step response curve, (t_p, y_p), (t_x, y_x), (t_y, y_y), and possibly (t_z, y_z). First, identify from the whole step response the system gain as $k = \frac{\Delta y}{\Delta u}$ and the delay h, if any. Choosing four points from the step response, the transfer function parameters are

$$\tau_1 = \frac{t_y - t_x}{\ln\left(\frac{y_{x\%} - 1}{y_{y\%} - 1}\right)} \tag{5.17}$$

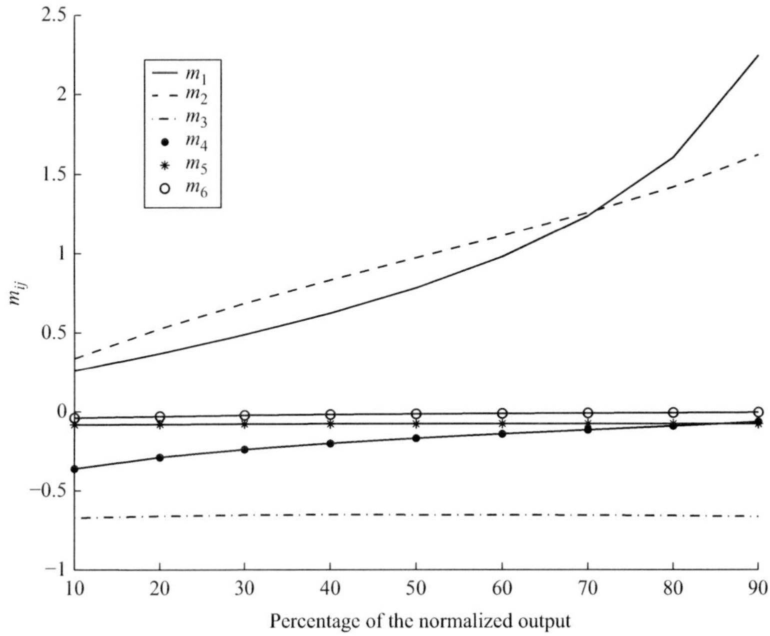

Figure 5.4 Parameters m_{ij}

$$b = 1 - \frac{1 - y_{p\%}}{e^{-t'_p}} \tag{5.18}$$

$$a = \frac{t'_z - (m_{1z} + m_{3z}\bar{b} + m_{5z}\bar{b}^2)}{m_{2z} + m_{4z}\bar{b} + m_{6z}\bar{b}^2} \tag{5.19}$$

We propose the following identification algorithm for simplicity and generality. The points considered for calculating dominant time constant τ_1 are $(t_{47\%}, y_{47\%})$ and $(t_{90\%}, y_{90\%})$. The constant a is calculated using again $(t_{47\%}, y_{47\%})$, thus just three points are required.

$$\begin{aligned} \tau_1 &= 0.60(t_{90\%} - t_{47\%}) \\ b &= 1 - \frac{1 - y_{p\%}}{e^{-t'_p}} \\ a &= \frac{t'_{47\%x} - n}{m} \end{aligned} \tag{5.20}$$

where the constants m and n are given in Table 5.1. Note that $m = m_{2x} + m_{4x}b + m_{6x}b^2$ and $n = m_{1x} + m_{3x}b + m_{5x}b^2$.

Table 5.1 Constants m and n for the proposed identification algorithm

b	m	n	b	m	n
0.1	0.9094	0.6905	2.1	0.4798	−0.8682
0.2	0.8910	0.6258	2.2	0.4549	−0.9607
0.3	0.8722	0.5597	2.3	0.4297	−1.0546
0.4	0.8532	0.4922	2.4	0.4042	−1.1500
0.5	0.8338	0.4233	2.5	0.3784	−1.2467
0.6	0.8141	0.3530	2.6	0.3522	−1.3448
0.7	0.7940	0.2814	2.7	0.3258	−1.4443
0.8	0.7737	0.2083	2.8	0.2990	−1.5452
0.9	0.7530	0.1338	2.9	0.2718	−1.6475
1.0	0.7320	0.0580	3.0	0.2444	−1.7512
1.1	0.7107	−0.0192	3.1	0.2166	−1.8563
1.2	0.6890	−0.0979	3.2	0.1886	−1.9627
1.3	0.6671	−0.1779	3.3	0.1602	−2.0706
1.4	0.6448	−0.2593	3.4	0.1314	−2.1799
1.5	0.6222	−0.3421	3.5	0.1024	−2.2905
1.6	0.5992	−0.4263	3.6	0.0730	−2.4025
1.7	0.5760	−0.5119	3.7	0.0433	−2.5160
1.8	0.5524	−0.5989	3.8	0.0133	−2.6308
1.9	0.5285	−0.6872	3.9	−0.0170	−2.7470
2.0	0.5043	−0.7770	4.0	−0.0477	−2.8646

Identification accuracy analysis

In order to gain insight on the identification accuracy we perform a prediction error evaluation of models identified for the admissible parameter range. Thus, we calculated the prediction error

$$S_2 = \sum_{k=1}^{\infty} (y(t_k) - y_m(t_k))^2 \tag{5.21}$$

where $y(t_k)$ is the output value of the exact step response and $y_m(t_k)$ is the output model value. In all cases the simulation time of step response has been 25 s, and the time increment $\Delta t = 0.025$ s. In Figure 5.5 the quadratic prediction error is shown as a function of a and b.

It can be seen that the error S_2 increases as a increases as expected, although for values of $b > 1.5$ the tendency becomes nearly erratic. However, it is difficult to find tendencies around the central zone, what implies that the algorithm makes use of important information because the residues show no tendency apart from the ones considered and expected above.

In Figure 5.6 we show the best and the worst identified step responses. The best model is given for parameters $a = 0.4$ and $b = 2.25$ yielding a prediction error value of $S_2 = 0.001579$. On the other hand the worst identified model is given for parameters $a = 0.2$ and $b = 4$ with prediction error $S_2 = 0.853486$.

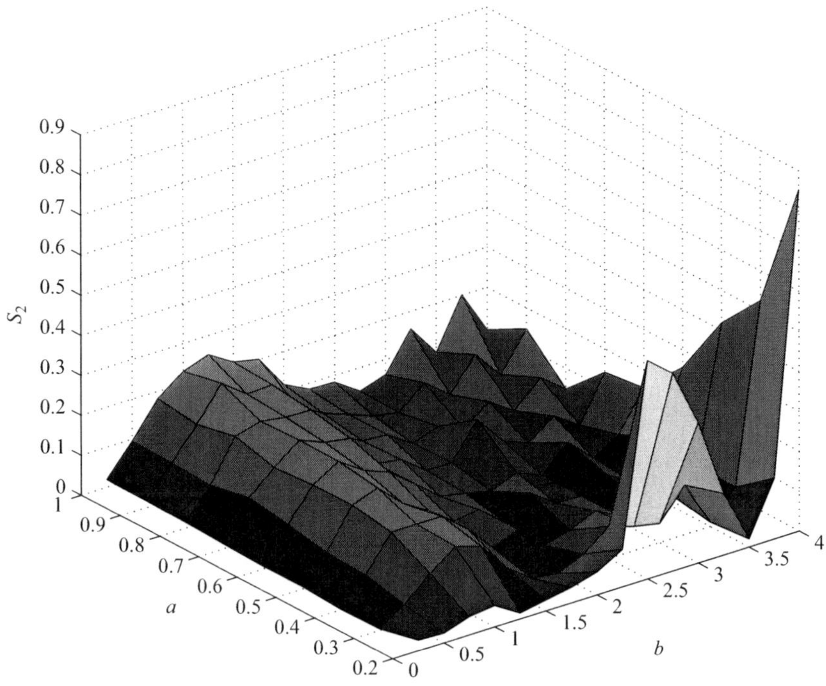

Figure 5.5 Prediction error S_2

5.3.4 Application examples

Numerical example

The proposed identification algorithm is applied to the identification of the plant given by $k = 1$, $b = 1$, $\tau_l = 1$, and time delay $h = 1$ for three different values of parameter $a = \{0.2, 0.35, 0.5\}$. The objective is to compare the algorithm performance with respect to distinct parameter a values. In Figure 5.7 the plant output is plotted against the identified plant output. The step input is of unit magnitude and applied at time zero. It can be seen that in all cases the responses are fairly similar.

Continuous stirred tank reactor

In this section a case-study example is provided. We consider the isothermal continuous stirred tank reactor (CSTR), as the one in Figure 5.8, where the isothermal series/parallel Van de Vusse reaction [4,12] is taking place. The reaction can be described by the following scheme:

$$A \xrightarrow{k_1} B \xrightarrow{k_2} C \qquad (5.22)$$

$$2A \xrightarrow{k_3} D$$

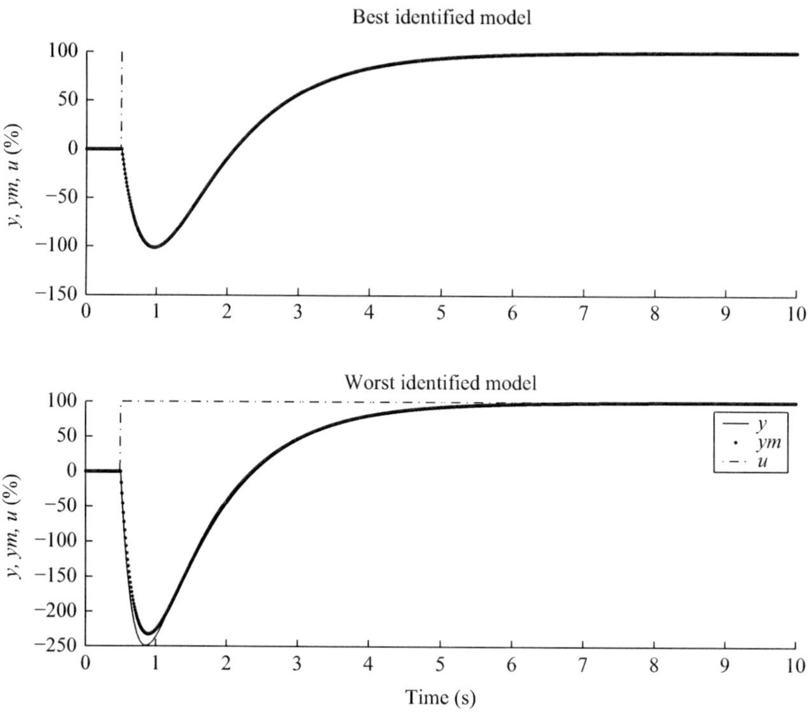

Figure 5.6 Best and worst identified models

Applying a mass balance, the system can be described by the following model:

$$\begin{aligned}\frac{dC_A(t)}{dt} &= \frac{F_r(t)}{V}[C_{Ai} - C_A(t)] - k_1 C_A(t) - k_3 C_A^2(t)\\ \frac{dC_B(t)}{dt} &= -\frac{F_r(t)}{V} C_B(t) + k_1 C_A(t) - k_2 C_B(t)\end{aligned} \tag{5.23}$$

where F_r is the feed flow rate of product A, V is the reactor volume, which is kept constant during the operation, C_A and C_B are the reactant concentrations in the reactor, and k_i $(i = 1, 2, 3)$ are the reaction rate constants for the three reactions.

In this case, the variables of interest are: the concentration of B in the reactor (C_B as the controlled variable), the flow through the reactor (F_r as the manipulated variable), and the concentration C_{Ai} of A in the feed flow (whose variation can be considered as the disturbance). The kinetic parameters are chosen to be $k_1 = 5/6\ \text{min}^{-1}$, $k_2 = 5/3\ \text{min}^{-1}$, and $k_3 = 1/6\ \text{l mol}^{-1}\ \text{min}^{-1}$. Also, it is assumed that the nominal concentration of A in the feed (C_{Ai}) is $10\ \text{mol l}^{-1}$ and the volume $V = 700$ l.

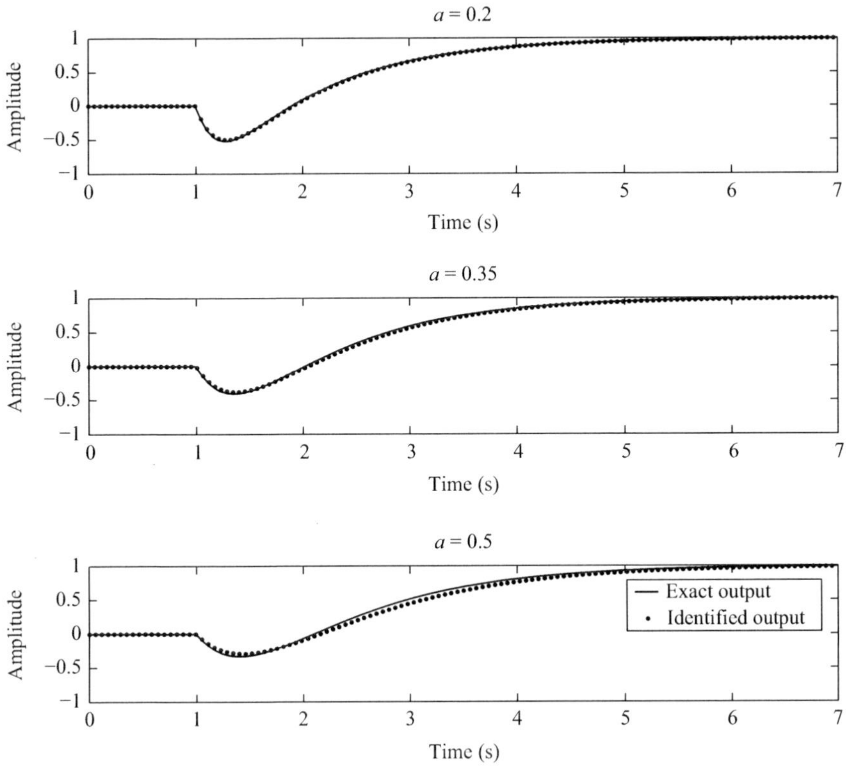

Figure 5.7 Example 1 – Step outputs' comparison

Using (5.23) and the parameters values, the characterization of the steady-state for the process can be obtained as it is shown in Figure 5.9, for three concentrations of C_{Ai}, where it is easy to see the non-linearity of the system.

Initially, the system is at the steady-state (therefore the operational point) with $C_{Ao} = 2.9175$ mol l^{-1} and $C_{Bo} = 1.10$ mol l^{-1}. It is assumed that changes in the set-point would be not bigger than 10% and the possible disturbance in C_{Ai} can variate around $\pm 10\%$.

Applying the proposed methodology according to (5.20), with a 10% change on the set-point, the identified process model is

$$G(s) = \frac{0.3199(-0.3520s + 1)}{(0.5619s + 1)(0.3086s + 1)} \tag{5.24}$$

Figure 5.10 shows the process output and the identified model (5.24) for a step change in the process input ($y_u(t)$). It is possible to see that the identification procedure has a very good accuracy, specially during the first time instants of the step response. The step response of the linearized model obtained on the operating point

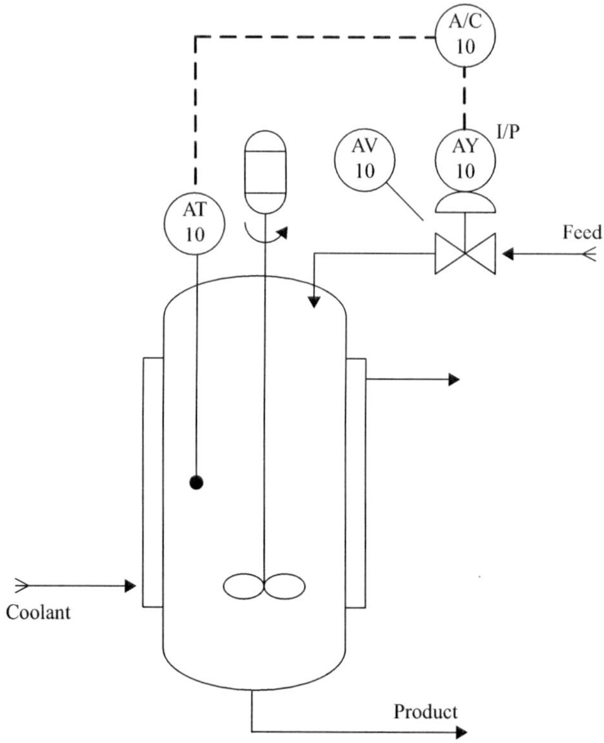

Figure 5.8 Example 2 – CSTR system

is also shown in Figure 5.10. The model obtained by the simple step response identification is better than the more complex linearized model obtained analytically. The applicability of the proposed algorithm to obtain accurate models for CSTR processes with second order inverse response is then justified.

5.4 Example 2: Reduced effective transfer function reduction for PID decentralized control

5.4.1 Problem statement

Processes in industrial settings are normally multivariable. Multivariable control systems may be designed using multivariable control structures [10]. However, multivariable controller design and tuning are far more complicated than of the well-established single loop control. As a result, one practical approach is to control multivariable processes by using single input/single output (SISO) control structures, that is decentralized control. Although decentralized control is less complex than multivariable control, even for industrial multivariable processes showing simple

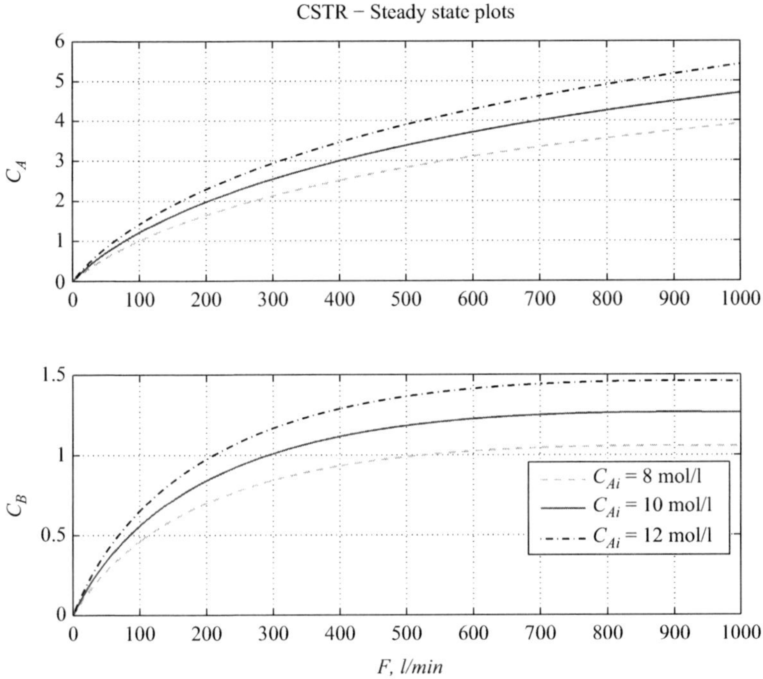

Figure 5.9 Example 2 – Steady-state characterization for the reactor

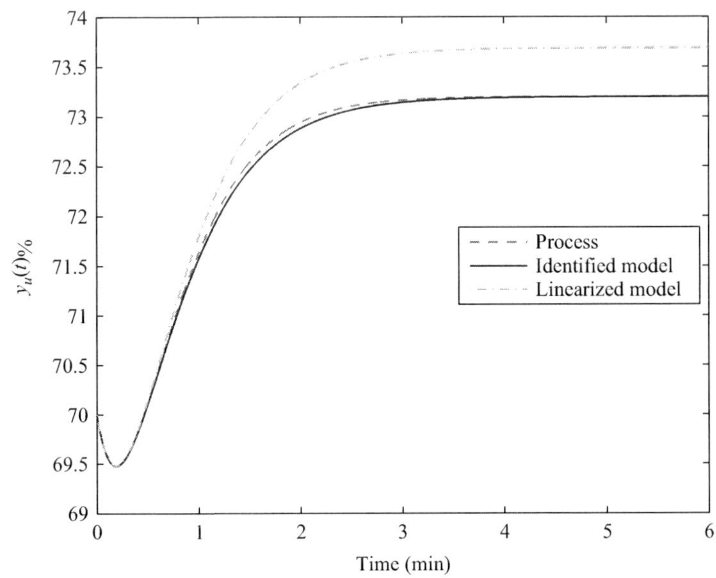

Figure 5.10 Example 2 – Reaction curve for process, linearized model, and identified model

dynamics, the design of decentralized controllers has to deal with complex model dynamics due to loop interaction. The complexity of resulting models dramatically limits the applicability of PID tuning rules existing in the literature [8], because they are based on simple models. For instance, consider the two inputs/two outputs (TITO) multivariable system shown in Figure 5.11. Furthermore, consider that transfer functions G_{11}, G_{12}, G_{21}, and G_{22} are modeled by first order plus dead time (FOPDT) models, that is the matrix transfer function is given by

$$G(s) = \begin{bmatrix} \dfrac{k_{11}e^{-h_{11}s}}{\tau_{11}s+1} & \dfrac{k_{12}e^{-h_{12}s}}{\tau_{12}s+1} \\ \dfrac{k_{21}e^{-h_{21}s}}{\tau_{21}s+1} & \dfrac{k_{22}e^{-h_{22}s}}{\tau_{22}s+1} \end{bmatrix} \tag{5.25}$$

In order to design decentralized controller C_{11} (see Figure 5.11), the transfer function between output Y_1 and input U_1, required to design decentralized controller C_{11}, is

$$G^e_{11} = \frac{Y_1}{U_1} = G_{11} - \frac{C_{22}G_{12}G_{21}}{1 + C_{22}G_{22}} \tag{5.26}$$

where for the sake of readability the complex variable s is not shown. Equation (5.26) is called effective transfer function (ETF) and it depends on the existing controller of closed loop pair C_{22}.

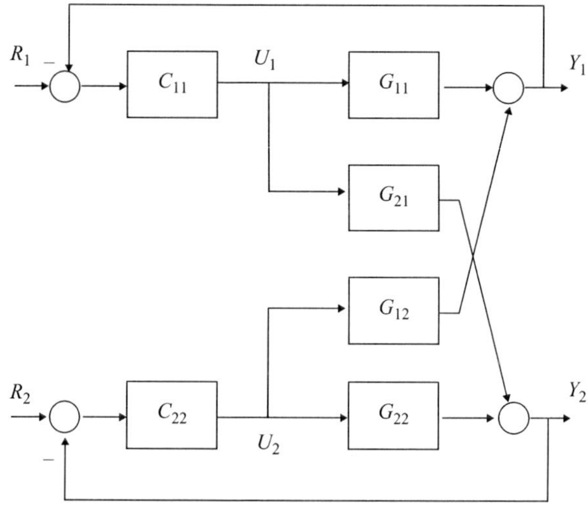

Figure 5.11 Decentralized control structure of a TITO system

For the sake of simplicity, the EFT dependence on existing controller C_{22} may be eliminated under the perfect control assumption, because in this case [10],

$$\frac{Y_2}{R_2} = \frac{C_{22}G_{22}}{1 + C_{22}G_{22}} = 1 \tag{5.27}$$

and the ETF can be simplified as follows:

$$G_{11}^{re} = G_{11} - \frac{G_{12}G_{21}}{G_{22}} \tag{5.28}$$

Equation (5.28) is known as reduced effective transfer function (RETF), and it depends on system transfer functions only.

Particularizing the RETF for the FOPDT TITO system, we have

$$G_{11}^{re} = \frac{k_{11}e^{-h_{11}s}}{(\tau_{11}s + 1)} - \frac{k_{12}k_{21}(\tau_{22}s + 1)e^{-(h_{12}+h_{21}-h_{22})s}}{k_{22}(\tau_{12}s + 1)(\tau_{21}s + 1)} \tag{5.29}$$

At the sight of transfer function (5.29) it is not clear if it can be accurately approximated by an FOPDT or an second order plus dead time (SOPDT) model, commonly used as baseline for PID control design. In fact transfer function (5.29) may even show inverse response behavior that cannot be captured by a minimum phase model. Dimensional analysis is applied for model order reduction of the RETF, characterizing the feasibility of approximating the RETF by means of FOPDT/SOPDT systems. Furthermore, in base of dimensionless numbers relations, it is possible to ascertain when transfer function (5.29) shows inverse response behavior, so care must be taken with the controller design procedure.

5.4.2 *Dimensionless representation of the reduced effective transfer function*

The first step of the approach is to represent the RETF (5.29) in dimensionless form, that is given by

$$\bar{G}_{11}^{re} = \frac{e^{-\bar{h}_{11}\bar{s}}}{(\bar{s} + 1)} - \bar{k}\frac{(\bar{\tau}_{22}\bar{s} + 1)e^{-\bar{h}_{\Sigma}\bar{s}}}{(\bar{\tau}_{12}\bar{s} + 1)(\bar{\tau}_{21}\bar{s} + 1)} \tag{5.30}$$

with the dimensionless numbers $\bar{G}_{11}^{re} = G_{11}^{re}/k_{11}$, $\bar{s} = \tau_{11}s$, $\bar{k} = (k_{12}k_{21})/(k_{11}k_{22})$, $\bar{\tau}_{12} = \tau_{12}/\tau_{11}$, $\bar{\tau}_{21} = \tau_{21}/\tau_{11}$, $\bar{\tau}_{22} = \tau_{22}/\tau_{11}$, $\bar{h}_{11} = h_{11}/\tau_{11}$, and $\bar{h}_{\Sigma} = (h_{12} + h_{21} - h_{22})/\tau_{11}$.

The benefit of dimensionless representation is two-fold. On the one hand we reduce the number of dependent parameters from eight dimensional parameters to six dimensionless parameters. Second, and more important, the dimensionless parameters enable the interpretation in terms of limiting values, what facilitates the analysis of particular cases. In the rest of the chapter we consider the following assumptions:

- All the previous dimensionless parameters are positive. The implication is stability of FOPDT transfer functions and that negative gain signs appear in pairs.
- For causality considerations, $\bar{h}_\Sigma \geq 0$, that is $h_{12} + h_{21} - h_{22} \geq 0$.
- We assume that $\bar{k} < 1$. By proper selection of input–output pairs, for instance consider pairing following the relative gain array (RGA) procedure, it follows that $\bar{k} < 1$.

5.4.3 *Inverse response analysis*

Given the dimensionless reduced effective transfer function (DRETF) (5.30), the first point of study is to characterize when (5.30) shows an inverse response behavior. In this case a nonminimum phase SOPDT model might be used. On the contrary, if the DRETF does not show inverse response, then an FOPDT or a minimum phase SOPDT model may be more appropriate to capture the fundamental dynamics of DRETF. In what follows, we provide conditions for inverse response behavior, first in the general case and, second, in some particular cases.

General case

Assume without loss of generality that $h_\Sigma > h_{11}$. On the contrary, if $h_\Sigma < h_{11}$, an analogue analysis may be performed. The case $h_\Sigma = h_{11}$ is analyzed in the next section as a particular case.

In case that $h_\Sigma > h_{11}$, the DRETF (5.30) may be rewritten as

$$\bar{G}_{11}^{re} = \left(\frac{1}{(\bar{s}+1)} - \bar{k}\frac{(\bar{\tau}_{22}\bar{s}+1)e^{-\bar{h}_x\bar{s}}}{(\bar{\tau}_{12}\bar{s}+1)(\bar{\tau}_{21}\bar{s}+1)} \right) e^{-\bar{h}_{11}\bar{s}} \tag{5.31}$$

where $\bar{h}_x := \bar{h}_\Sigma - \bar{h}_{11} > 0$.

Transfer function (5.31) shows inverse response if the difference of transfer functions between parentheses has zeros on the right half plane (RHP), that is positive zeros [10]. Note that the time delay $\bar{h}_{11}$ has no influence on the existence of inverse response.

In order to characterize the inverse response property of transfer function (5.31), we approximate the time delay $\bar{h}_x$ by a first order Padé approximation, yielding

$$\bar{G}_{11}^{re} = \left(\frac{1}{(\bar{s}+1)} - \bar{k}\frac{(\bar{\tau}_{22}\bar{s}+1)(\frac{-\bar{h}_x}{2}\bar{s}+1)}{(\bar{\tau}_{12}\bar{s}+1)(\bar{\tau}_{21}\bar{s}+1)(\frac{\bar{h}_x}{2}\bar{s}+1)} \right) e^{-\bar{T}_{11}\bar{s}} \tag{5.32}$$

which can be rewritten as

$$\bar{G}_{11}^{re} = \frac{a\bar{s}^3 + b\bar{s}^2 + c\bar{s} + d}{(\bar{s}+1)(\bar{\tau}_{12}\bar{s}+1)(\bar{\tau}_{21}\bar{s}+1)(\frac{\bar{h}_x}{2}\bar{s}+1)} e^{-\bar{h}_{11}\bar{s}} \tag{5.33}$$

where the third order polynomial of the numerator $P(\bar{s}) := a\bar{s}^3 + b\bar{s}^2 + c\bar{s} + d$ has parameters defined as

$$\begin{aligned}
a &= (\bar{\tau}_{12}\bar{\tau}_{21} + \bar{k}\bar{\tau}_{22})(\bar{h}_x/2) \\
b &= \bar{\tau}_{12}\bar{\tau}_{21} + \frac{\bar{h}_x}{2}(\bar{\tau}_{12} + \bar{\tau}_{21}) - \bar{k}\big(\bar{\tau}_{22} - (1+\bar{\tau}_{22})\bar{h}_x/2\big) \\
c &= \bar{\tau}_{12} + \bar{\tau}_{21} + \frac{\bar{h}_x}{2} - \bar{k}\left(\bar{\tau}_{22} - \frac{\bar{h}_x}{2} + 1\right) \\
d &= 1 - \bar{k}
\end{aligned}$$

Moreover, we define the discriminant Δ of $P(\bar{s})$ as $\Delta = 18abcd - 4b^3d + b^2c^2 - 4ac^3 - 27a^2d^2$ [15].

Theorem 5.1. *Assume that $P(\bar{s})$ has only real valued zeros (i.e. $\Delta \geq 0$). Transfer function (5.33) is minimum phase (i.e. all zeros are left half plane (LHP) zeros) if and only if*

$$\bar{\tau}_{22} < \min\left\{ \frac{\bar{\tau}_{12}\bar{\tau}_{21} + \frac{\bar{h}_x}{2} + \bar{k}\frac{\bar{h}_x}{2}}{\bar{k}(1 - \frac{\bar{h}_x}{2})},\ \frac{\bar{\tau}_{12} + \bar{\tau}_{21} + \frac{\bar{h}_x}{2}(1+\bar{k})}{\bar{k}} - 1 \right\} \tag{5.34}$$

Proof. By Descartes' rule of signs we know that (i) the maximum number of positive zeros is equal to the changes of sign of $P(\bar{s})$, and (ii) the minimum number of negative zeros is equal to the changes of sign of $P(-\bar{s})$.

Note that term a is always greater than zero because we are adding positive values. Moreover, d is also greater than zero because, by proper pairing, $\bar{k} < 1$. As a result, the number of sign changes is determined by parameters b and c. If $b > 0$ and $c > 0$ then the number of sign changes of $P(\bar{s})$ is zero, thus there are no positive zeros (i.e. no inverse response). In fact, in this case, the number of sign changes of $P(-\bar{s})$ is equal to 3, the number of negatives zeros.

If any of the parameters b or c, or both, is negative, then the number of sign changes of $P(\bar{s})$ is 2, whereas the number of sign changes of $P(-\bar{s})$ is 1. Two possibilities are then open: there are two positive zeros and one negative zero; or there are two complex conjugate zeros and one negative zero. The existence of complex zeros is discarded by assumption $\Delta \geq 0$. Therefore, there are two positive real zeros and the system is in nonminimum phase.

Finally, $b > 0$ is equivalent to

$$\bar{\tau}_{22} < \frac{\bar{\tau}_{12}\bar{\tau}_{21} + \frac{\bar{h}_x}{2} + \bar{k}\frac{\bar{h}_x}{2}}{\bar{k}\left(1 - \frac{\bar{h}_x}{2}\right)} \tag{5.35}$$

and $c > 0$ is equivalent to

$$\bar{\tau}_{22} < \frac{\bar{\tau}_{12} + \bar{\tau}_{21} + \frac{\bar{h}_{\Sigma}}{2}(1 + \bar{k})}{\bar{k}} - 1 \tag{5.36}$$

and the result follows. □

Time delay factorization

Assuming that $h_{\Sigma} = h_{11}$, both delays in (5.30) may be factored, then the RETF may be written as

$$\bar{G}_{11}^{re} = \left(\frac{1}{(\bar{s} + 1)} - \bar{k} \frac{(\bar{\tau}_{22}\bar{s} + 1)}{(\bar{\tau}_{12}\bar{s} + 1)(\bar{\tau}_{21}\bar{s} + 1)} \right) e^{-\bar{h}_{11}\bar{s}} \tag{5.37}$$

In this case, the time delay has no influence on the inverse response manifestation of the DRETF and the inverse response behavior of transfer function (5.37) may be analyzed equivalently by transfer function

$$\bar{G}_{11}^{re} = \frac{(\bar{\tau}_{12}\bar{s} + 1)(\bar{\tau}_{21}\bar{s} + 1) - \bar{k}(\bar{\tau}_{22}\bar{s} + 1)(\bar{s} + 1)}{(\bar{s} + 1)(\bar{\tau}_{12}\bar{s} + 1)(\bar{\tau}_{21}\bar{s} + 1)} e^{-\bar{h}_{11}\bar{s}} \tag{5.38}$$

The inverse response behavior occurs when transfer function (5.38) has, at least, one right half plane zero (RHP). The zeros of the numerator polynomial are given by

$$a\bar{s}^2 + b\bar{s} + c = 0 \tag{5.39}$$

with $a = \bar{\tau}_{12}\bar{\tau}_{21} - \bar{k}\bar{\tau}_{22}$, $b = \bar{\tau}_{12} + \bar{\tau}_{21} - \bar{k}(\bar{\tau}_{22} + 1)$, and $c = (1 - \bar{k})$.

The zeros of polynomial (5.39) are given by $\bar{s} = (-b \pm \sqrt{b^2 - 4ac})/(2a)$. For real zeros, by imposing the condition $-b + \sqrt{b^2 - 4ac} < 0$, we assure that the zeros are LHP zeros and no inverse response is possible. By arranging and squaring previous condition we have $b^2 - 4ac < b^2$, yielding the condition $ac > 0$. This is written in terms of dimensionless numbers as

$$(\bar{\tau}_{12}\bar{\tau}_{21} - \bar{k}\bar{\tau}_{22})(1 - \bar{k}) > 0 \tag{5.40}$$

which is a necessary and sufficient condition to discard nonminimum phase behavior in (5.37). As by assumption $\bar{k} < 1$, condition (5.40) is simplified to

$$\bar{k} < \frac{\bar{\tau}_{12}\bar{\tau}_{21}}{\bar{\tau}_{22}} \tag{5.41}$$

In case that poles are complex conjugates, the condition for LHP zeros is that $b > 0$, that is

$$\bar{k} < \frac{\bar{\tau}_{12} + \bar{\tau}_{21}}{\bar{\tau}_{22} + 1} \tag{5.42}$$

Pole-zero cancelation

The case of pole-zero cancelation occurs when $\bar{\tau}_{12} = \bar{\tau}_{22}$ (or $\bar{\tau}_{21} = \bar{\tau}_{22}$) yields to further simplifications. In this case transfer function (5.31) may be reduced to

$$\bar{G}_{11}^{re} = \left(\frac{1}{\bar{s}+1} - \bar{k}\frac{\left(\frac{-\bar{h}_x}{2}\bar{s}+1\right)}{(\bar{\tau}_{21}\bar{s}+1)\left(\left(\frac{\bar{h}_x}{2}\bar{s}+1\right)\right)} \right) e^{-\bar{h}_{11}} \tag{5.43}$$

The inverse response behavior occurs when transfer function (5.43) has an RHP. The zeros of the numerator polynomial are given by

$$a\bar{s}^2 + \bar{s} + c = 0 \tag{5.44}$$

with $a = (\bar{\tau}_{21} + \bar{k})\frac{\bar{h}_x}{2}$, $b = \bar{\tau}_{21} + \frac{\bar{h}_x}{2} - \bar{k}(1 + \frac{\bar{h}_x}{2})$, and $c = (1 - \bar{k})$.

By solving equation (5.44) we arrive at the condition for minimum phase for real zeros, given by

$$(\bar{\tau}_{21} + \bar{k})\frac{\bar{h}_x}{2}(1 - \bar{k}) > 0 \tag{5.45}$$

which by assumption we have $0 < \bar{k} < 1$, and the non-minimum phase behavior is discarded for any RETF.

For complex zeros the condition to discard inverse response is that $b > 0$, which yields

$$\bar{k} < \frac{\bar{\tau}_{21} + \frac{\bar{h}_x}{2}}{\frac{\bar{h}_x}{2} + 1} \tag{5.46}$$

Time delay factorization and pole-zero cancelation

The case of time delay factorization and pole-zero cancelation yields to further simplifications. In this case transfer function (5.31) may be reduced to

$$\bar{G}_{11}^{re} = \left(\frac{1}{\bar{s}+1} - \bar{k}\frac{1}{\bar{\tau}_{21}\bar{s}+1} \right) \tag{5.47}$$

The difference of two first order systems yields an inverse response if $\frac{1}{\bar{\tau}_{21}} > \frac{1}{\bar{k}} > 1$ [5]. By assumption, we have $0 < \bar{k} < 1$, it is always true that $\frac{1}{\bar{k}} > 1$, thus, there is inverse response behavior if and only if

$$\bar{k} > \bar{\tau}_{21} \tag{5.48}$$

Table 5.2 Summary of no inverse response conditions for each case considered

Assumptions	**No inverse response if...**
$\Delta \geq 0$	$\bar{\tau}_{22} < \min \left\{ \frac{\bar{\tau}_{12}\bar{\tau}_{21} + \frac{\bar{h}_{\Sigma}}{2} + \bar{k}\frac{\bar{h}_{\Sigma}}{2}}{\bar{k}(1-\frac{\bar{h}_{\Sigma}}{2})}, \frac{\bar{\tau}_{12}+\bar{\tau}_{21}+\frac{\bar{h}_{\Sigma}}{2}(1+\bar{k})}{\bar{k}} - 1 \right\}$
$\bar{h}_{11} = \bar{h}_{\Sigma}$, $\Delta \geq 0$	$\bar{k} < \frac{\bar{\tau}_{12}\bar{\tau}_{21}}{\bar{\tau}_{22}}$
$\bar{h}_{11} = \bar{h}_{\Sigma}$, $\Delta < 0$	$\bar{k} < \frac{\bar{\tau}_{12}+\bar{\tau}_{21}}{\bar{\tau}_{22}+1}$
$\bar{\tau}_{12} = \bar{\tau}_{22}$, $\Delta \geq 0$	No inverse response
(or $\bar{\tau}_{21} = \bar{\tau}_{22}$), $\Delta < 0$	$\bar{k} < \frac{\bar{\tau}_{21}+\frac{\bar{h}_{\Sigma}}{2}}{\frac{\bar{h}_{\Sigma}}{2}+1}$
$\bar{h}_{11} = \bar{h}_{\Sigma}$ and $\bar{\tau}_{12} = \bar{\tau}_{22}$ (or $\bar{\tau}_{21} = \bar{\tau}_{22}$)	$\bar{k} < \bar{\tau}_{21}$

Summary of inverse response analysis

In Table 5.2 we present a summary of inverse response conditions for each case considered. Recall that the discriminant for the general case is defined as $\Delta = 18abcd - 4b^3d + b^2c^2 - 4ac^3 - 27a^2d^2$, whereas the discriminant for the rest of cases is defined as $\Delta = b^2 - 4ac$.

5.4.4 Reduced order model: general case

One general approach to obtain reduced order models of the RETF is to perform Maclaurin approximations, as presented in Reference 14. The idea is to expand the RETF in its Maclaurin series as follows:

$$\bar{G}_{11}^{re} = a_{11} + b_{11}\bar{s} + c_{11}\bar{s}^2 + O(\bar{s}^3) \tag{5.49}$$

where the polynomial coefficients are

$$\begin{aligned} a_{11} &= \bar{G}_{11}^{re}(0) \\ b_{11} &= \left.\frac{d\bar{G}_{11}^{re}(\bar{s})}{d\bar{s}}\right|_{\bar{s}=0} \\ c_{11} &= \left.\frac{1}{2}\frac{d^2\bar{G}_{11}^{re}(\bar{s})}{d\bar{s}^2}\right|_{\bar{s}=0} \end{aligned}$$

Next, we obtain the analytic polynomial coefficients of the Maclaurin series for the dimensionless FOPDT model

$$\bar{G}(s) = \frac{k}{k_{11}}\frac{e^{-\bar{h}\bar{s}}}{\bar{\tau}\bar{s} + 1} \tag{5.50}$$

that is given by

$$\bar{G}^{re} = \frac{k}{k_{11}} - \frac{k}{k_{11}}(\bar{\tau} + \bar{h})\bar{s} + \frac{k}{k_{11}}\left(\frac{1}{2}\bar{h}^2 + (\bar{h} + \bar{\tau})\bar{\tau}\right)\bar{s}^2 + O(\bar{s}^3) \tag{5.51}$$

For SOPDT processes, the same procedure may be repeated. However, the SOPDT parameters are not easily obtainable as before [14]. Section 5.4.5 shows how SOPDT models may be approximated without requiring the Maclaurin approximation.

5.4.5 Reduced order model: particular cases

At the sight of dimensionless number values, we may obtain reduced order models in a straightforward manner. In fact, if some dimensionless numbers are small or there is certain relationship among them, simpler models may be easily obtained.

Dimensionless gain

In case that $\bar{k} \ll 1$, an approximation to the dimensionless reduced effective transfer function is simply

$$\frac{1}{\bar{s} + 1}e^{-\bar{h}_{11}\bar{s}} \tag{5.52}$$

As $\bar{k} = (k_{12}k_{21})/(k_{11}k_{22})$, the condition $\bar{k} \ll 1$ implies that the cross gains are small and we have low interaction. As a result we may consider the problem as not coupled.

Slow model

Previous model may be improved in the case of two time scales. We have two time scales if $\bar{\tau}_{12} \ll 1$ and $\bar{\tau}_{21} \ll 1$. In this case we may disregard fast model dynamics. The DRETF (5.29) is given by the slow model with FOPDT transfer function given by

$$\frac{(1 - \bar{k})}{\bar{s} + 1}e^{-h_{11}\bar{s}} \tag{5.53}$$

However, note that when $\bar{\tau}_{12} \ll 1$ and $\bar{\tau}_{21} \ll 1$, it might be possible to perform time scale decoupling [9], and the original loop pairing might not be the most adequate, despite RGA pairing.

Time delay factorization and pole-zero cancelation

Recall that under the assumption $h_\Sigma = h_{11}$, equivalently $h_{11} + h_{22} = h_{12} + h_{21}$, the delay can be factored, so the dimensionless reduced effective transfer function is given by (5.37). In this case the exact dimensionless reduced effective transfer function is given by the following transfer function:

$$\bar{G}_{11}^{re} = \frac{(\bar{\tau}_{12}\bar{s}+1)(\bar{\tau}_{21}\bar{s}+1) - \bar{k}(\bar{\tau}_{22}\bar{s}+1)(\bar{s}+1)}{(\bar{s}+1)(\bar{\tau}_{12}\bar{s}+1)(\bar{\tau}_{21}\bar{s}+1)} e^{-\bar{h}_{11}\bar{s}} \tag{5.54}$$

that is too complex to be used for control design with commonly used PID tuning rules.

The case of pole-zero cancelation by $\bar{\tau}_{12} = \bar{\tau}_{22}$ (or $\bar{\tau}_{21} = \bar{\tau}_{22}$) yields to further simplifications. In this case transfer function (5.54) may be reduced to

$$\bar{G}_{11}^{re} = \frac{(\bar{\tau}_{12} - \bar{k})\bar{s} + (1-\bar{k})}{(\bar{s}+1)(\bar{\tau}_{12}\bar{s}+1)} e^{-\bar{h}_{11}\bar{s}} \tag{5.55}$$

Transfer function (5.55) is exactly modeled by second order transfer function. We have the following cases:

- $(\bar{\tau}_{12} - \bar{k}) < 0$. In this case a non-minimum phase SOPDT model is able to exactly represent the dimensionless reduced effective transfer function. There is inverse response.
- $(\bar{\tau}_{12} - \bar{k}) > 0$. In this case a minimum phase with LHP zero SOPDT model is able to exactly represent the dimensionless reduced effective transfer function.
- $(\bar{\tau}_{12} - \bar{k}) = 0$. In this case an SOPDT model is able to exactly represent the dimensionless reduced effective transfer function.

Summary of model order reduction

In Table 5.3 we present a summary of reduced order models.

Table 5.3 Summary of model order reduction for each case considered

Assumptions	**Model**
$\bar{k} \ll 1$	$\frac{1}{\bar{s}+1} e^{-\bar{h}_{11}\bar{s}}$
$\bar{\tau}_{12} \ll 1$ and $\bar{\tau}_{21} \ll 1$	$\frac{(1-\bar{k})}{\bar{s}+1} e^{-\bar{h}_{11}\bar{s}}$
$\bar{h}_{11} = \bar{h}_{\Sigma}$ and $\bar{\tau}_{12} = \bar{\tau}_{22}$ (or $\bar{\tau}_{21} = \bar{\tau}_{22}$)	$\frac{(\bar{\tau}_{12} - k)\bar{s} + (1-k)}{(\bar{s}+1)(\bar{\tau}_{12}\bar{s}+1)} e^{-\bar{h}_{11}\bar{s}}$
$\bar{h}_{11} = \bar{h}_{\Sigma}$ and $\bar{\tau}_{12} = \bar{\tau}_{22}$ (or $\bar{\tau}_{21} = \bar{\tau}_{22}$) and $\bar{\tau}_{12} = 1$	$\frac{(1-\bar{k})}{\bar{s}+1} e^{-\bar{h}_{11}\bar{s}}$

5.4.6 Application examples

Vinante and Luyben (VL) column

Consider the Vinante and Luyben distillation column [13]

$$G(s) = \begin{bmatrix} \dfrac{-2.2e^{-s}}{7s+1} & \dfrac{1.3e^{-0.3s}}{7s+1} \\ \dfrac{-2.8e^{-1.8s}}{9.5s+1} & \dfrac{4.3e^{-0.35s}}{9.2s+1} \end{bmatrix} \tag{5.56}$$

The RETF dimensionless parameters are given in Table 5.4.

Table 5.4 Vinante and Luyben DRETF dimensionless numbers

$\bar{k}$	$\bar{\tau}_{12}$	$\bar{\tau}_{21}$	$\bar{\tau}_{22}$	$\bar{h}_{11}$	$\bar{h}_{\Sigma}$
0.38	1	1.35	1.31	0.14	0.25

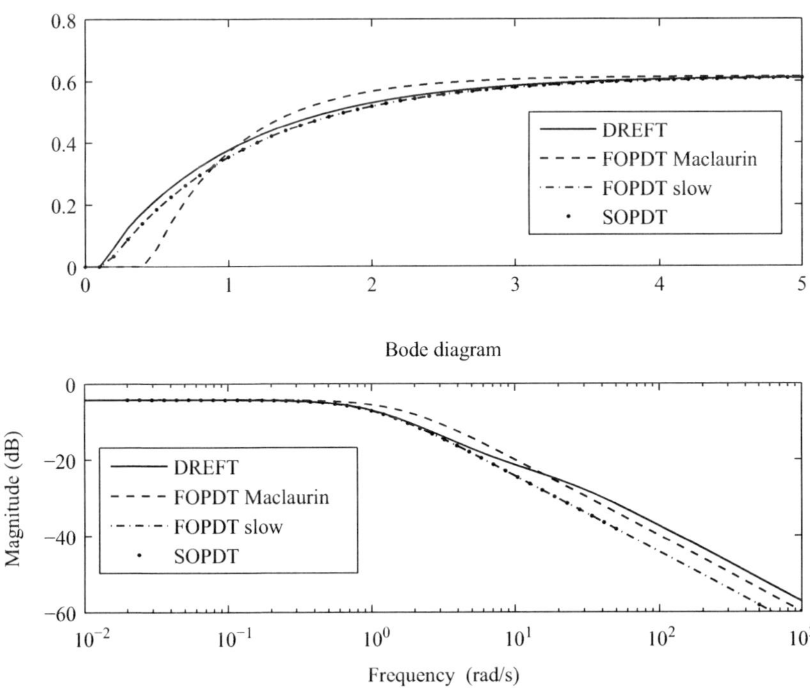

Figure 5.12 Step response and Bode magnitude plot of reduced models of Vinante and Luyben DRETF

The DRETF has no inverse response behavior because the discriminant is $\Delta = 0.07 > 0$, and applying Theorem 5.1 (condition (5.34)), we have

$$1.31 < \min\{1.46, 5.31\} \tag{5.57}$$

which discards inverse response behavior.

Note that $\bar{\tau}_{21} \approx \bar{\tau}_{22}$, thus by the pole-zero cancelation case, we can also discard non-minimum phase behavior using condition (5.46), that yields

$$0.38 < 1.33 \tag{5.58}$$

which is true and discards inverse response behavior of DRETF.

As inverse response behavior is discarded, two FOPDT models and one minimum phase SOPDT model are proposed: first, the 'FOPDT Maclaurin' approximation that requires no assumptions; second, the 'FOPDT Slow' model obtained assuming that $\bar{k} \ll 1$ and given by (5.53); third, the 'SOPDT' model derived by assuming (i) $\bar{\tau}_{21} = \bar{\tau}_{22}$, and (ii) $\bar{h}_{11} = \bar{h}_{\Sigma}$. Figure 5.12 shows the step responses and the magnitude Bode diagram of the models considered.

Polymerization reactor

Consider now the polymerization reactor [2]

$$G(s) = \begin{bmatrix} \dfrac{22.89e^{-0.2s}}{4.572s+1} & \dfrac{-11.64e^{-0.4s}}{1.807s+1} \\ \dfrac{4.689e^{-0.2s}}{2.174s+1} & \dfrac{5.80e^{-0.4s}}{1.801s+1} \end{bmatrix} \tag{5.59}$$

The RETF dimensionless parameters are given in Table 5.5. Note that $\bar{\tau}_{12} \approx \bar{\tau}_{22}$. Furthermore, we have $\bar{h}_{11} = \bar{h}_{\Sigma}$, so we consider the time delay factorization and pole-zero cancelation case. Note also that in this case $\bar{k} < 0$, hence the results must be particularized to this case by realizing that no inverse response is possible because two first order transfer functions are being added in (5.47).

The same three models as previous examples are used and the results can be seen in Figure 5.13. Note that in this case, the SOPDT model is equal to the DRETF.

Table 5.5 Polymerization reactor DRETF dimensionless numbers

$\bar{k}$	$\bar{\tau}_{12}$	$\bar{\tau}_{21}$	$\bar{\tau}_{22}$	$\bar{h}_{11}$	$\bar{h}_{\Sigma}$
−0.41	0.39	0.47	0.39	0.0437	0.0437

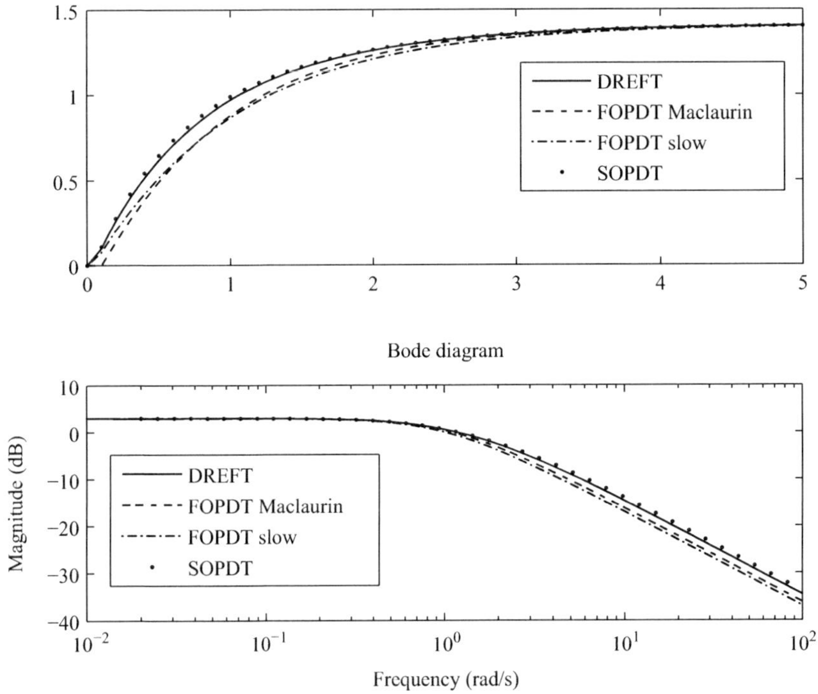

Figure 5.13 Step response and Bode magnitude plot of reduced order models of Polymerization Reactor DRETF

References

[1] K.J. Åström and T. Hägglund. *PID Controllers: Theory, Design and Tuning.* Instrument Society of America, 1995

[2] J. Garrido, F. Vázquez, and F. Morilla. 'Centralized inverted decoupling for TITO processes.' In *15th IEEE International Conference on Emerging Technologies and Factory Automation*, Bilbao, Spain, 2010

[3] H.P. Huang, M.W. Lee, and C.L. Chen. 'A system of procedures for identification of simple models using transient step response.' *Industrial and Engineering Chemistry Research*, 40(8):1903–1915, 2001

[4] C. Kravaris and P. Daoutidis. 'Nonlinear state feedback control of second order nonminimum-phase nonlinear systems.' *Computers & Chemical Engineering*, 14(4–5):439–449, 1990

[5] K. Linoya and R.J. Altpeter. 'Inverse response in process control.' *Industrial and Engineering Chemistry Research*, 54(7):39–43, 1962

[6] W.L. Luyben. 'Identification and tuning of integral processes with deadtime and inverse response.' *Industrial and Engineering Chemistry Research*, 42:3030–3035, 2003

[7] T.E. Marlin. *Process Control. Designing Process and Control Systems for Dynamic Performance*. McGraw-Hill, 2000

[8] A. O'Dwyer. *Handbook of PI and PID Controller Tuning Rules*. Imperial College Press, 2006

[9] B.A. Ogunnaike and W. Harmor Ray. *Process Dynamics, Modelling and Control*. Oxford University Press, 1994

[10] S. Skogestad and I. Postlethwaite. *Multivariable Feedback Control. Analysis and Design*. Wiley, 1996

[11] A. Srivastava and A.K. Verma. 'Identification of integrating processes with deadtime and inverse response.' *Industrial and Engineering Chemistry Research*, 46:8270–8272, 2007

[12] J.G. Van de Vusse. Plug-flow type reactor versus tank reactor. *Chemical Engineering Science*, 19:964, 1964

[13] C.D. Vinante and W.L. Luyben. Experimental studies of distillation decoupling. *Kemin Teollisuusmaalaus*, 1972

[14] T.N.L. Vu and M. Lee. Independent design of multi-loop pi/pid controllers for interacting multivariable processes. *Journal of Process Control*, 20:922–933, 2010

[15] E.W. Weisstein. *CRC Concise Encyclopedia of Mathematics*. Chapman & Hall/CRC, 2003

Chapter 6

Homogeneity of PID tuning rules

with A. Ibeas, C. Pedret and S. Alcántara

6.1 Introduction

We have seen that any physical quantity, apart from its value, has a unit that measures its physical dimension. Physical dimensions and units, like length (m), mass (kg), and time (s), relate mathematics and the physical world by physical equations. Valid physical equations describing physical system behavior are dimensionally homogeneous. In the same way, valid controller equations, the controller being a physical system relating measured outputs with control actions, are also dimensionally homogeneous. In contrast, the controller tuning rule, that is the rule providing the controller parameters as a function of model parameters, is just a mathematical relationship and, a priori, there is not any physical constraint requiring dimensional homogeneity.

In fact, homogeneous tuning rules form a subset of the possible tuning rules, as nonhomogeneous tuning rules are discarded. As a result, is the homogeneity property of a tuning rule limiting the achievable performance? Could be the case that nonhomogeneous tuning rules would provide a better closed loop performance? For instance, for PID controllers, dimensionally homogeneous tuning rules and nonhomogeneous tuning rules can be found in Reference 2.

In the present chapter we demonstrate that tuning rules derived from minimization of indexes depending on closed loop transfer functions yield homogeneous and nonhomogeneous tuning rules, depending on the closed loop functions considered in the minimization index.

However, when index minimization yields nonhomogeneous tuning rules, the index can be transformed by dimensional weighting factors, recovering homogeneity property of the tuning rule. As a result, no optimality is lost by using homogeneous tuning rules, and the prevalence of dimensional homogeneous tuning rules is explained.

However, note that when dimensional weighting factors are used in the minimization index, the dimensionless weights depend on the plant parameters value instead of being a constant factor. As a result, homogeneous tuning rules have a property lacking on nonhomogeneous tuning rules, that is dimensionally homogeneous tuning rules preserve dimensional similarity.

In this way, closed loops of similar plants tuned with homogeneous tuning rules are also similar, which implies that their closed loop responses are related by gain and time scaling factors. In other words, closed loop responses of dimensionally similar plants are equal except for gain and time scaling when dimensionally similar tuning rules are used. Consider for instance that experimental results show a 5% overshoot on the step response. If a dimensionally homogeneous tuning rule is used, we can expect the same 5% overshoot on the prototyped airplane. In this case equality follows because overshoot is a dimensionless parameter.

It can be concluded that homogeneous tuning rules are more advantageous than nonhomogeneous ones because, although both tuning procedures may yield the same tuning performance, homogeneous tuning rules also preserve dimensional similarity, whereas nonhomogeneous tuning rules do not.

6.2 Homogeneous PID tuning rules

6.2.1 Dimensionless controller parameters

Consider the following ideal form of PID controller:

$$C(s) = k_C\left(1 + \frac{1}{\tau_I s} + \tau_D s\right) \tag{6.1}$$

with the controller parameters k_C, τ_I, and τ_D.

The dimensionless representation of the controller is given by following steps (i)–(iii) presented in Chapter 3, and in step (iv), $C(\bar{s})$ is divided by k_C, thus

$$\bar{C}(\bar{s}) := \frac{C(\bar{s})}{k_C} = \left(1 + \frac{1}{\bar{\tau}_I \bar{s}} + \bar{\tau}_D \bar{s}\right) \tag{6.2}$$

with dimensionless numbers defined as $\bar{\tau}_I := \frac{\tau_I}{\tau_{p1}}$, $\bar{\tau}_D := \frac{\tau_D}{\tau_{p1}}$, and $\bar{s} := \tau_{p1}s$.

The relation between the dimensional and dimensionless controller is given by

$$C(s, \theta_C) = k_C \bar{C}(\bar{s}, \Pi'_C) \tag{6.3}$$

with $\theta_C = \{k_C, \tau_I, \tau_D\}$ and $\Pi'_C = \{\bar{\tau}_I, \bar{\tau}_D\}$.

Note that the dimensionless controller numbers Π_C have been obtained by the use of plant parameter τ_{p1}. This is mandatory in order to have the same dimensionless variable $\bar{s}$ as the plant, which allows the frequency scale compatibility between dimensionless plant and dimensionless controller.

6.2.2 Homogeneous tuning rules characterization

In this section we characterize the function of tuning rules that are dimensionally homogeneous, so we can compare homogeneous and nonhomogeneous tuning rules. We assume that the controller parameters depend on the plant parameters $\theta_G := \{k, h, \tau_{p1}, \ldots, \tau_{pn}, \tau_{c1}, \ldots, \tau_{cm}, l\}$ and, possibly, on closed loop specifications θ_λ. Examples of θ_λ are the closed loop time constant λ and the sensitivity function peak M_s. The tuning rule f_i, $i = 1, \ldots, 3$, sets the controller parameters dependence on model parameters θ_G and tuning parameters θ_λ

$$k_C = f_1(\theta_G, \theta_\lambda) \tag{6.4}$$

$$\tau_I = f_2(\theta_G, \theta_\lambda) \tag{6.5}$$

$$\tau_D = f_3(\theta_G, \theta_\lambda) \tag{6.6}$$

We assume that the tuning-related parameters dimensions are either time dimension $[\theta_\lambda] = T$ or dimensionless $[\theta_\lambda] = 1$. For instance $[\lambda] = T$ and $[M_s] = 1$. In this way, parameters of set θ_G and θ_λ in (6.4)–(6.6) have only two distinct dimensions, the gain dimension K and the time dimension T. Moreover, by definition $[G(s)] := [Y]/[U]$, where Y is output dimension and U is input dimension, and $[C(s)] := [U]/[Y]$. As a result, $[G(s)C(s)] = 1$, that is dimensionless. Recall that $[G(s)] = [k\tau_{p1}^l]$ and $[C(s)] = [k_C]$, thus $[k_C k\tau_{p1}^l] = 1$, that is, the controller gain k_C is made dimensionless multiplying by $k\tau_{p1}^l$.

Definition 6.1. A tuning rule is homogeneous if it can be written in dimensionless form as follows:

$$k_C k\tau_{p1}^l = \Phi_1(\Pi_G, \Pi_\lambda) \tag{6.7}$$

$$\frac{\tau_I}{\tau_{p1}} = \Phi_2(\Pi_G, \Pi_\lambda) \tag{6.8}$$

$$\frac{\tau_D}{\tau_{p1}} = \Phi_3(\Pi_G, \Pi_\lambda) \tag{6.9}$$

Recall that the controller parameters dimensionless representation is based on the homogeneity assumption. Although any valid physically based equation is homogeneous, a tuning rule may be nonhomogeneous.

The following two examples show homogeneous and nonhomogeneous tuning rules.

Example 6.1. The Ziegler–Nichols tuning rule [7] is

$$k_C = 1.2\tau/kh$$

$$\tau_I = 2h$$

$$\tau_D = \frac{1}{2}h$$

It can be written in dimensionless form as

$$k_C k = 1.2\left(\frac{h}{\tau}\right)^{-1}$$

$$\frac{\tau_I}{\tau} = \frac{2h}{\tau}$$

$$\frac{\tau_I}{\tau} = \frac{1}{2}\frac{h}{\tau}$$

Example 6.2. On the contrary, the Calender *et al.* tuning rule [2, p. 78] is

$$k_C = \frac{1.066}{kh}$$

$$\tau_I = 1.418h$$

$$\tau_D = 0.47h$$

and $k_C k$ is no longer dimensionless but has inverse time dimension. As a result the tuning rule is not homogeneous and cannot be written in dimensionless form.

Example 6.3. Table 6.1 presents the dimensionless form of generic PID tuning rules for a first-order plus dead-time (FOPDT) process and for a second-order plus dead-time (SOPDT) process. Moreover, the tuning rule for the SOPDT process also considers the tuning parameter λ defined as the closed loop time constant. As a result dimensionless tuning rules for FOPDT process depend only on one dimensionless parameter, whereas dimensionless tuning rules for SOPDT process depend on three dimensionless parameters. Note also that in both cases the plant gain only affects the controller gain. The rest of controller parameters do not depend on the plant gain.

Table 6.1 Homogeneous tuning rules for two distinct plants. Tuning rules for the first-order plus dead-time (FOPDT) processes depend only on one dimensionless parameter h/τ. On the other hand, homogeneous tuning rules for second-order plus dead-time (SOPDT) processes, with closed loop time constant tuning parameter λ, depend on three dimensionless parameters

Dimensionally homogeneous tuning rules functions		
Π_C	$FOPDT = \dfrac{ke^{-hs}}{(\tau s+1)}$	$SOPDT = \dfrac{ke^{-hs}}{(\tau_1 s+1)(\tau_2 s+1)}$
$k\tau_{p1}^{l} k_C$	$\Phi_1\left(\dfrac{h}{\tau}\right)$	$\Phi_1\left(\dfrac{\tau_2}{\tau_1}, \dfrac{h}{\tau_1}, \dfrac{\lambda}{\tau_1}\right)$
$\dfrac{\tau_I}{\tau}$	$\Phi_2\left(\dfrac{h}{\tau}\right)$	$\Phi_2\left(\dfrac{\tau_2}{\tau_1}, \dfrac{h}{\tau_1}, \dfrac{\lambda}{\tau_1}\right)$
$\dfrac{\tau_D}{\tau}$	$\Phi_3\left(\dfrac{h}{\tau}\right)$	$\Phi_3\left(\dfrac{\tau_2}{\tau_1}, \dfrac{h}{\tau_1}, \dfrac{\lambda}{\tau_1}\right)$

6.2.3 Dimensionless controller representation with homogeneous tuning rules

Dimensional and dimensionless controllers tuned with homogeneous tuning rules are related by

$$C(s, \theta_C) = k_C \bar{C}(\bar{s}, \Pi_G, \Pi_\lambda) \tag{6.10}$$

The relationship follows by recalling that the general dimensionless controller form is related to its dimensional counterpart by $C(s, \theta_C) = k_C \bar{C}(\bar{s}, \Pi_C^{'})$. Moreover, as a result of tuning rule homogeneity, we have $\Pi_C = \Phi_i(\Pi_G, \Pi_\lambda)$, $i = 1, \ldots, 3$, and the result follows.

6.3 Closed loop transfer functions

Once tuning rule homogeneity has been defined, we characterize the closed loop transfer functions of homogeneous tuning rules. The objective is to obtain the dimensionless form of closed loop transfer functions. Thereafter, it will be possible to ascertain the homogeneity of tuning rules obtained by minimization of indexes depending on closed loop transfer functions.

6.3.1 Loop transfer function

The following lemma establishes that the open loop transfer function $G(s)C(s)$ obtained from homogeneous tuning rules is a function of the plant dimensionless numbers and of the specifications dimensionless numbers, if any.

Lemma 6.1. *Given a transfer function $G(s)$ and a PID controller $C(s)$ with parameters k_C, τ_I, and τ_D, given by an arbitrary homogeneous tuning rule defined in dimensionless form by*

$$k_C k \tau_{p1}^{l} = \Phi_1 (\Pi_G, \Pi_\lambda) \tag{6.11}$$

$$\tau_I / \tau_{p1} = \Phi_2 (\Pi_G, \Pi_\lambda) \tag{6.12}$$

$$\tau_D / \tau_{p1} = \Phi_3 (\Pi_G, \Pi_\lambda) \tag{6.13}$$

the loop transfer function $L(s) = G(s)C(s)$, which depends on the variable s and the parameters θ_G and θ_C, that is $L = L(s, \theta_G, \theta_C)$, can be represented as a dimensionless function depending on the normalized variable $\bar{s} = \tau_{p1} s$ and the dimensionless parameters Π_G and Π_λ, thus

$$L = L(\bar{s}, \Pi_G, \Pi_\lambda) \tag{6.14}$$

Proof. The relation between dimensional and dimensionless model is $G(s) = k\tau_{p1}^{l}\bar{G}(\bar{s}, \Pi_G)$, and the relation between dimensional and dimensionless controller is $C(s) = k_C\bar{C}(\bar{s}, \Pi_G, \Pi_\lambda)$. It follows that

$$\begin{aligned} G(s)C(s) &= k_C k \tau_{p1}^{l} \bar{G}(\bar{s}, \Pi_G)\bar{C}(\bar{s}, \Pi_G, \Pi_\lambda) \\ &= \Phi_1(\Pi_G, \Pi_\lambda)\bar{G}(\bar{s}, \Pi_G)\bar{C}(\bar{s}, \Pi_G, \Pi_\lambda) \\ &= L(\bar{s}, \Pi_G, \Pi_\lambda) \end{aligned} \tag{6.15}$$

□

In case of nonhomogeneous tuning rules the open loop transfer function $L(s)$ is a function of at least one dimensional parameter θ_G and/or θ_λ.

Example 6.4. Consider an FOPDT model. First of all, recall that a general FOPDT transfer function can be rewritten by means of the dimensionless number h/τ

$$G(s) = G(\bar{s}) = \frac{k}{(\bar{s}+1)} e^{-\frac{h}{\tau}\bar{s}} \tag{6.16}$$

The ideal PID controller $C(s)$ can be written as

$$C(s) = k_C \frac{\tau_I \tau_D s^2 + \tau_I s + 1}{\tau_I s} \tag{6.17}$$

For any homogeneous tuning rule we have

$$k_C = k^{-1}\Phi_1\left(\frac{h}{\tau}\right) \tag{6.18}$$

$$\tau_I = \tau\Phi_2\left(\frac{h}{\tau}\right) \tag{6.19}$$

$$\tau_D = \tau\Phi_3\left(\frac{h}{\tau}\right) \tag{6.20}$$

Substituting the tuning rules structure of (6.18)–(6.20) into the PID transfer function (6.17) we obtain

$$C(s) = k^{-1}\Phi_1\left(\frac{h}{\tau}\right) \frac{\tau^2\Phi_2\left(\frac{h}{\tau}\right)\Phi_3\left(\frac{h}{\tau}\right)s^2 + \tau\Phi_2\left(\frac{h}{\tau}\right)s + 1}{\tau\Phi_2\left(\frac{h}{\tau}\right)s} \tag{6.21}$$

taking into account that $\bar{s} = \tau s$, (6.21) yields

$$C(s) = C(\bar{s}) = k^{-1}\Phi_1\left(\frac{h}{\tau}\right) \frac{\Phi_2\left(\frac{h}{\tau}\right)\Phi_3\left(\frac{h}{\tau}\right)\bar{s}^2 + \Phi_2\left(\frac{h}{\tau}\right)\bar{s} + 1}{\Phi_2\left(\frac{h}{\tau}\right)\bar{s}} \tag{6.22}$$

then multiplying $G(s)$ by $C(s)$ we obtain the open loop transfer function $L(s)$ as

$$L(s) = L(\bar{s}) = \frac{\Phi_1\left(\frac{h}{\tau}\right)}{\Phi_2\left(\frac{h}{\tau}\right)} \frac{\Phi_2\left(\frac{h}{\tau}\right)\Phi_3\left(\frac{h}{\tau}\right)\bar{s}^2 + \Phi_2\left(\frac{h}{\tau}\right)\bar{s} + 1}{(\bar{s}+1)\bar{s}} e^{-(h/\tau)\bar{s}} \tag{6.23}$$

which can also be simplified to

$$L(\bar{s}) = \Phi_1\left(\frac{h}{\tau}\right) \frac{\Phi_3\left(\frac{h}{\tau}\right)\bar{s}^2 + \bar{s} + \Phi_2^{-1}\left(\frac{h}{\tau}\right)}{(\bar{s}+1)\bar{s}} e^{-(h/\tau)\bar{s}} \tag{6.24}$$

Thus, the loop transfer function $L(s)$ is a function of the dimensionless variable h/τ.

6.3.2 Dimensionless closed loop transfer functions

Consider the control loop given in Figure 6.1. The reference r, input disturbance d_u, and output disturbance d_y are the exogenous signals, whereas the error e, the control action u, and the output y are the internal signals [1]. By definition internal signals and exogenous signals are related by

$$\begin{aligned} e &= r - y \\ u &= d_u + Ce \\ y &= d_y + Gu \end{aligned}$$

which can be written in matrix form as

$$\begin{bmatrix} 1 & 0 & 1 \\ -C & 1 & 0 \\ 0 & -G & 1 \end{bmatrix} \begin{bmatrix} e \\ u \\ y \end{bmatrix} = \begin{bmatrix} r \\ d_u \\ d_y \end{bmatrix} \tag{6.25}$$

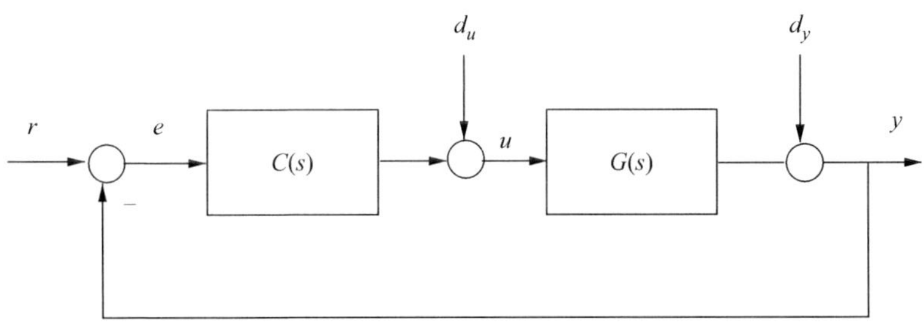

Figure 6.1 Control structure. Any relationship between exogenous and internal signals is given by any of the following four closed loop transfer functions, the sensitivity $S(s)$, the complementary sensitivity $T(s)$, the control sensitivity $S_C(s)$, and the plant sensitivity $S_G(s)$. The closed loop is completely determined by these four transfer functions

We are interested in describing all the internal signals as a function of exogenous signals, thus

$$\begin{bmatrix} e \\ u \\ y \end{bmatrix} = \begin{bmatrix} 1 & 0 & 1 \\ -C & 1 & 0 \\ 0 & -G & 1 \end{bmatrix}^{-1} \begin{bmatrix} r \\ d_u \\ d_y \end{bmatrix} \tag{6.26}$$

which results in

$$\begin{bmatrix} e \\ u \\ y \end{bmatrix} = \frac{1}{1+GC} \begin{bmatrix} 1 & -G & -1 \\ C & 1 & -C \\ GC & G & 1 \end{bmatrix} \begin{bmatrix} r \\ d_u \\ d_y \end{bmatrix} \tag{6.27}$$

As a result, the relation between internal signals and exogenous signals is completely determined by the following four transfer functions that determine the closed loop behavior:

(i) $S = (1+GC)^{-1} = \frac{e}{r} = \frac{e}{d_y} = \frac{u}{d_u} = \frac{y}{d_y}$
(ii) $T = GC(1+GC)^{-1} = \frac{y}{r}$
(iii) $S_C = C(1+GC)^{-1} = \frac{u}{r} = \frac{-u}{d_y}$
(iv) $S_G = G(1+GC)^{-1} = \frac{-e}{d_u} = \frac{y}{d_u}$

S is the sensitivity transfer function, T is the complementary sensitivity transfer function, S_C is the control sensitivity transfer function, and S_G is the plant sensitivity transfer function.

Recall that $\bar{G} = \bar{G}(\bar{s}, \Pi_G)$ and $\bar{C} = \bar{C}(\bar{s}, \Pi_C)$ are the dimensionless model and dimensionless controller, respectively, hence $L(\bar{s}) = \bar{G}(\bar{s})\bar{C}(\bar{s})$. Dividing by τ_{p1} and performing the change of variable $\bar{s} \rightarrow \tau_{p1}s$, the previous closed loop transfer functions are $S = (1+k_Ck\tau_{p1}^l L)^{-1}$, $T = k_Ck\tau_{p1}^l L(1+k_Ck\tau_{p1}^l L)^{-1}$, $S_C = k_C\bar{C}(1+k_Ck\tau_{p1}^l L)^{-1}$, and $S_G = k\tau_{p1}^l \bar{G}(1+k_Ck\tau_{p1}^l L)^{-1}$.

In case of homogeneous tuning rules it follows that $\Pi_C = \Phi_i(\Pi_G, \Pi_\lambda)$, $i = 1,\ldots,3$, and in particular $k_Ck\tau_{p1}^l = \Phi_1(\Pi_G, \Pi_\lambda)$. As a result, the previous closed loop transfer functions show the following parameters dependence: $S(s) = S(\bar{s}, \Pi_G, \Pi_\lambda)$, $T(s) = T(\bar{s}, \Pi_G, \Pi_\lambda)$, $S_C(s) = k_C\bar{S}_C(\bar{s}, \Pi_G, \Pi_\lambda)$, and $S_G(s) = k\tau_{p1}^l \bar{S}_G(\bar{s}, \Pi_G, \Pi_\lambda)$

The implication is that $S(s) = S(\bar{s})$ and $T(s) = T(\bar{s})$ because they are already dimensionless. In fact, transfer functions $S(\bar{s})$ and $T(\bar{s})$ relate signals with the same dimensions (reference and output), thus they are dimensionless. On the contrary this is not the case for transfer functions $S_C(\bar{s})$ and $S_G(\bar{s})$, which have dimensional gain because $S_C(\bar{s})$ and $S_G(\bar{s})$ relate signals with, in general, distinct dimensions (output and input).

6.4 Optimality of homogeneous tuning rules

In this section we investigate the conditions of the scalar index $I = I(S(s), T(s), S_C(s), S_G(s))$ which minimization yields homogeneous tuning rules. An example of a scalar index is the minimization of the mixed sensitivity $\min_C ||w_S S;\ w_T T;\ w_u S_c||_\infty$ [4, p. 63]. As a result we explain not only the prevalence of homogeneous tuning rules but also their equivalent optimality properties when compared with nonhomogeneous tuning rules.

Lemma 6.2. *Any tuning rule that minimizes a scalar index that depends on the loop transfer function $L(s)$ only, that is $I = I(S(s))$, $I = I(T(s))$, or $I = I(S(s), T(s))$, is homogeneous.*

Proof. The transfer function $L(\bar{s})$ is a function of $L(\bar{s}) = L(\bar{s}, \Pi_G, \Pi_C)$. As a result transfer functions $S(\bar{s})$ and $T(\bar{s})$, or any combination, can be expressed in dimensionless form as functions of the dimensionless numbers Π_G and Π_C, that is

$$I = I(\bar{s}, \Pi_G, \Pi_C) \tag{6.28}$$

Assume that the optimal solution is given by Π_C^*. From (6.28), Π_C^* depends only on Π_G, hence the optimal tuning rules are homogeneous. □

Lemma 6.3. *Any tuning rule that minimizes a scalar index that depends on the control sensitivity $I = I(S_C(s))$ or on the plant sensitivity $I = I(S_G(s))$ is homogeneous.*

Proof. First note that $S_G(\bar{s}) = k\bar{S}_G(\bar{s}, \Pi_G, \Pi_C)$, and the constant term k does not affect the minimization result. As a result the optimum Π_C^* is a function of Π_G only. In the case of $S_C(\bar{s}) = k_C\bar{S}_C(\bar{s}, \Pi_G, \Pi_C)$, multiplying by $k\tau_{p1}^l$ does not affect the optimal solution because $k\tau_{p1}^l$ is also a constant, hence $k\tau_{p1}^l S_C(\bar{s}) = k\tau_{p1}^l k_C\bar{S}_C(\bar{s}, \Pi_G, \Pi_C)$, and $k\tau_{p1}^l k_C \in \Pi_C$. Thus, the minimization problem is homogeneous. □

Lemma 6.4. *Tuning rules that minimize any scalar index that combines $S_C(s)$ and $S_G(s)$ or any loop transfer function $L(s)$ with $S_C(s)$ and/or $S_G(s)$ are not homogeneous.*

Proof. Assume that we want to minimize the index $I = I(S_C, S_G)$, that is

$$\min_{\Pi_C}\ I(k_C\bar{S}_C(\bar{s}, \Pi_G, \Pi_C), k\tau_{p1}^l\bar{S}_G(\bar{s}, \Pi_G, \Pi_C)) \tag{6.29}$$

Multiplying (6.29) by constant $k\tau_{p1}^l$ yields an equivalent minimization problem because the minimum is not modified by a constant, thus

$$\min_{\Pi_C}\ k\tau_{p1}^l I(k_C\bar{S}_C(\bar{s}, \Pi_G, \Pi_C), k\tau_{p1}^l\bar{S}_G(\bar{s}, \Pi_G, \Pi_C)) \tag{6.30}$$

Moreover, by the property of homogeneous functions [5] we have (6.30) equal to

$$\min_{\Pi_C} I((k\tau_{p1}^l)^w k_C \bar{S}_C(\bar{s}, \Pi_G, \Pi_C), (k\tau_{p1}^l)^w k\tau_{p1}^l \bar{S}_G(\bar{s}, \Pi_G, \Pi_C)) \tag{6.31}$$

If $w = 1$ then

$$\min_{\Pi_C} I(\bar{S}_C(\bar{s}, \Pi_G, \Pi_C), k^2\tau_{p1}^{2l} \bar{S}_G(\bar{s}, \Pi_G, \Pi_C)) \tag{6.32}$$

which is not homogeneous.

If $w \neq 1$ then $(k\tau_{p1}^l)^w k_C \notin \Pi_C$, $(k\tau_{p1}^l)^w k\tau_{p1}^l \notin \Pi_G$, and $((k\tau_{p1}^l)^w k_C)/((k\tau_{p1}^l)^w k\tau_{p1}^l) \notin \Pi_C$. Thus, the minimization problem is not homogeneous. □

6.4.1 Weighting factors

In the preceding section the homogeneity theorems were discussed without the inclusion of weighting factors when constructing the indices to be minimized. However, the inclusion of weighting factors in the minimization indexes permits to suit the optimization problems to be solved. The weighting factors used have a twofold purpose: on the one hand, the numerical value that adjusts the function to be minimized, and on the other hand, the units of the weighting factor that can set either a homogeneous or a nonhomogeneous minimization problem.

Lemma 6.5. *Any tuning rule that minimizes a homogeneous scalar index given by $I = I(L, \alpha S_C, \beta S_G)$, with α, β weighting factors with dimensions $[\alpha] = KT^l$ and $[\beta] = (KT^l)^{-1}$, is given by*

$$\Pi_C = \Gamma(\Pi_G, \bar{\alpha}, \bar{\beta}) \tag{6.33}$$

with $\bar{\alpha} = \alpha/(k\tau_{p1}^l)$ and $\bar{\beta} = \beta k\tau_{p1}^l$.

Proof. After scaling $\bar{s} = \tau_{p1}s$ it follows that $[L(\bar{s})] = 1$, that is dimensionless. In the same way $[\alpha S_C(\bar{s})] = 1$ and $[\beta S_G(\bar{s})] = 1$, hence

$$I = I(L(\bar{s}, \Pi_G, \Pi_C), \alpha k_C \bar{S}_C(\bar{s}, \Pi_G, \Pi_C), \beta k\tau_{p1}^l \bar{S}_G(\bar{s}, \Pi_G, \Pi_C)) \tag{6.34}$$

Now defining $k_C k\tau_{p1}^l = \Pi_{CK}$ and recalling that $\Pi_{CK} \in \Pi_C$, (6.34) can be written as

$$I = I(L(\bar{s}, \Pi_G, \Pi_C), \frac{\alpha}{k\tau_{p1}^l} \Pi_{CK} \bar{S}_C(\bar{s}, \Pi_G, \Pi_C), \beta k\tau_{p1}^l \bar{S}_G(\bar{s}, \Pi_G, \Pi_C)) \tag{6.35}$$

In (6.35) the dimensionless controller parameters Π_C depend on Π_G, $\bar{\alpha} = \alpha/(k\tau_{p1}^l)$, and $\bar{\beta} = \beta k\tau_{p1}^l$, which are dimensionless parameters too. □

The theorem implication is that for distinct values of $k\tau_{p1}^{l}$, the homogeneous tuning rules are the solution of minimization problems where the dimensional weighting factors α and β are changed in order to keep their dimensionless counterparts constant.

The result explains the prevalence of homogeneous tuning rules in the bibliography as can be seen in Reference 2. First the most commonly used minimization indexes yield homogeneous tuning rules by themselves. Second, any minimization index can yield a homogeneous tuning rule by proper selection of weighting factors units. Moreover, the minimization of homogeneous indexes can be achieved by means of homogeneous tuning rules, as a result no optimality is lost by using homogeneous tuning rules instead of the more general nonhomogeneous ones. In the next section we derive the implications of using homogeneous tuning rules.

6.5 Homogeneous and nonhomogeneous tuning rules

For a single plant, homogeneous tuning can solve any minimization problem. As a result, nonhomogeneous tuning rules provide no advantage. In fact any minimization problem can be cast in dimensionless form by merely adding the appropriate dimensional weighting factors. So we can conclude that there is no limitation on the achievable performance if homogeneous tuning rules are used.

However, homogeneous and nonhomogeneous tuning rules are far from being equivalent. The difference between homogeneity and nonhomogeneity arises when investigating the tuning rules properties over sets of plants that share the similarity property. Consider for instance the set of LTI plants $G(s)$ of fixed but nevertheless arbitrary structure as can be seen in Figure 6.2. The set is partitioned into subsets of dimensionally similar plants. Each subset, due to similarity, has a single dimensionless representation that is marked by a black dot.

Now consider the closed loop transfer functions obtained from each plant by any homogeneous tuning rule. The homogeneous tuning rule maps the subset of dimensionally similar plants to the subset of closed loop transfer function $G_{cl}(s) = \{L, S, T, S_C, S_G\}$ that is also similar, in such a way that yields a single dimensionless closed loop representation. On the contrary, nonhomogeneous tuning rules map dimensionally similar subsets $G(s)$ into dimensionally nonsimilar sets on $G_{cl}(s)$. As a result the closed loop similarity is lost.

The use of homogeneous tuning rules is useful when, for instance, control systems are designed on the basis of models because experimental results can be easily transferred to the prototype. This is not the case if nonhomogeneous tuning rules are used. We remark that no optimality is lost in the use of homogeneous tuning rules. The previous discussion is formalized in the following lemmas.

Lemma 6.6. *Any homogeneous tuning rule maps the set of dimensionally similar plants defined by its dimensionless number* Π_G *into a set of dimensionally similar closed loop systems, defined by the closed loop transfer functions* $G_{cl}(s) = \{L, S, T, S_C, S_G\}$.

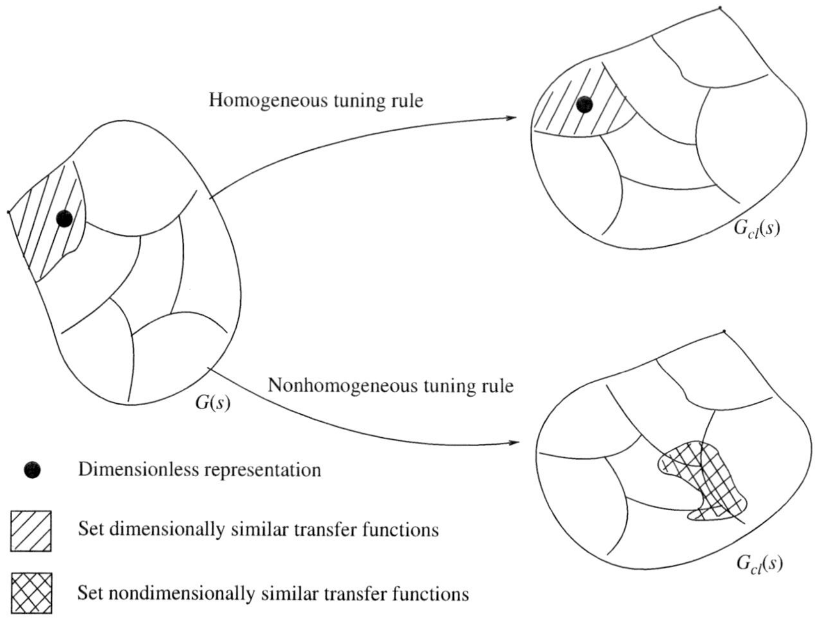

Figure 6.2 Open loop–closed loop transformation. The set of open loop systems $G(s)$ (left) is partitioned into dimensionally similar subsets. The black dot indicates the dimensionless representation. The set of closed loop systems $G_{cl} = \{S(s), T(s), S_C(s), S_G(s)\}$ (right) is also partitioned into dimensionally similar subsets. Homogeneous tuning rules preserve dimensional similarity. An infinite set of dimensionally similar plants is transformed by homogeneous tuning rules into closed loops transfer functions, which are also dimensionally similar. If the tuning rule is nonhomogeneous the dimensionally similar plants are mapped into closed loop transfer functions, which are no longer dimensionally similar, hence there is no single dimensionless representation for the closed loop transfer functions

Proof. The closed loop transfer functions are $S = S(\Pi_G, \Pi_C)$, $T = T(\Pi_G, \Pi_C)$, $S_C = k_C S_C(\Pi_G, \Pi_C)$, and $S_G = kS_G(\Pi_G, \Pi_C)$. Now, when a tuning rule is homogeneous $\Pi_C = \Phi(\Pi_G, \Pi_\lambda)$, we can write $S = S(\Pi_G, \Pi_\lambda)$, $T = T(\Pi_G, \Pi_\lambda)$, $S_C = k_C S_C(\Pi_G, \Pi_\lambda)$, and $S_G = kS_G(\Pi_G, \Pi_\lambda)$. As a result the set of dimensionally similar plants Π_G has the same set of dimensionally similar closed loop transfer functions. □

Lemma 6.7. *A nonhomogeneous tuning rule maps the set of dimensionally similar plants defined by its dimensionless number Π_G into closed loop sets $G_{cl}(s) = \{L, S, T, S_C, S_G\}$, which are not dimensionally similar among them.*

Proof. If a tuning rule is nonhomogeneous then $\Pi_C = \Phi(\theta_G, \theta_\lambda)$. As a result $S = S(\theta_G, \theta_\lambda)$, $T = T(\theta_G, \theta_\lambda)$, $S_C = k_C S_C(\theta_G, \theta_\lambda)$, and $S_G = k S_G(\theta_G, \theta_\lambda)$. As a result the set of dimensionally similar plants Π_G has not the same set of dimensionally similar closed loop transfer functions. □

Example 6.5. Consider the unstable time-delayed second-order transfer function given by

$$G(s) = \frac{k}{(1-\tau_1 s)(1+\tau_2 s)} e^{-hs} \tag{6.36}$$

We define two dimensionally similar transfer functions, the model $G_m(s)$ and the prototype $G_p(s)$, related by the gain scaling factor S_k and the time scaling factor S_t as follows:

$$\begin{aligned} k_p &= S_k k_m \\ \tau_{1p} &= S_t \tau_{1m} \\ \tau_{2p} &= S_t \tau_{2m} \\ h_p &= S_t h_m \end{aligned}$$

We investigate the closed loop properties of the model and the prototype transfer functions by PID control tuned with the homogeneous tuning rule of Rotstein and Lewin [3] and the nonhomogeneous tuning rule of Wang and Jin [2,6]. In both cases the tuning rule requires the desired closed loop time constant λ. The tuning rule of Rotstein and Lewin [3] is

$$\begin{aligned} k_C &= \frac{\tau_1\left(\lambda\left(\frac{\lambda}{\tau_1}+2\right)+\tau_2\right)}{\lambda^2 k} \\ \tau_I &= \lambda\left(\frac{\lambda}{\tau_1}+2\right)+\tau_2 \\ \tau_D &= \frac{\lambda\left(\frac{\lambda}{\tau_1}+2\right)\tau_2}{\lambda\left(\frac{\lambda}{\tau_1}+2\right)+\tau_2} \end{aligned} \tag{6.37}$$

By inspection, the tuning rule is homogeneous.

The tuning rule of Wang and Jin [2,6] is

$$\begin{aligned} k_C &= \frac{k\tau_1(\tau_1-h)(\tau_2+h)}{(\lambda+h)^3} \\ \tau_I &= \frac{\lambda^3+3h\lambda^2+3(\tau_1\tau_2+h\tau_1-h\tau_2)\lambda+h\tau_2(\tau_1-h)}{(\tau_1-h)(\tau_2+h)} \\ \tau_D &= \frac{(\tau_2+h-\tau_1)\lambda^3+3\tau_1\tau_2\lambda(\lambda+h)+\tau_1\tau_2 h^2}{\tau_I(\tau_1-h)(\tau_2+h)} \end{aligned} \tag{6.38}$$

In this case the tuning rule is no longer homogeneous because the controller gain has the same units as the plant gain. However, the integral and derivative time are homogeneous relations.

In what follows $k_m = 1$, $\tau_{1m} = 5$, $\tau_{2m} = 3$, and $h_m = 0.4$. The particular measurement units are not important. The selected closed loop time constant of the model is $\lambda_m = 2.5$. Model and prototype are related by the gain scaling factor $S_k = 1.2$ and the time scaling factor $S_t = 2$. The closed loop step responses for the homogeneous tuning of Rotstein and Lewin can be seen in Figure 6.3. As expected both model and prototype step responses are dimensionally similar. This is shown by temporally stretching the model step response and seeing that it equals the prototype step response.

Now consider the nonhomogeneous tuning rule of Wang and Jin applied to the same model and prototype. Figure 6.4 shows the closed loop step response. In this

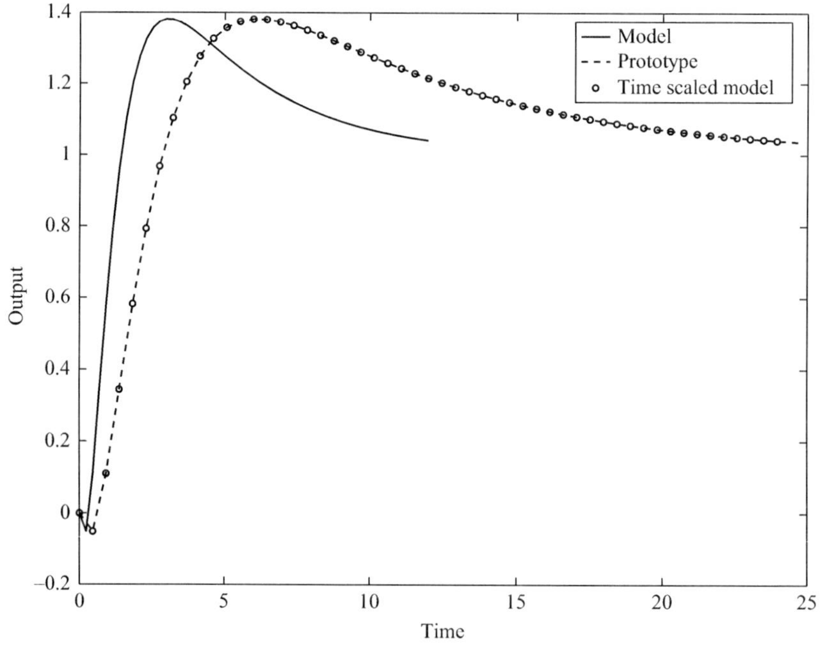

Figure 6.3 Homogeneous tuning rule step responses. When two distinct but dimensionally similar plants, the prototype $G_p(s)$ and the model $G_m(s)$, are controlled by proportional integral derivative (PID) control tuned with the Rotstein and Lewin homogeneous tuning rule, the closed loop behavior is also dimensionally similar. Consider the closed loop step responses for the model (-) and the prototype (- -). Visually both figures share the same shape. In fact, if the model response is time scaled (o) by scaling factor S_t it is equal to the prototype time response. Thus both closed loops are similar

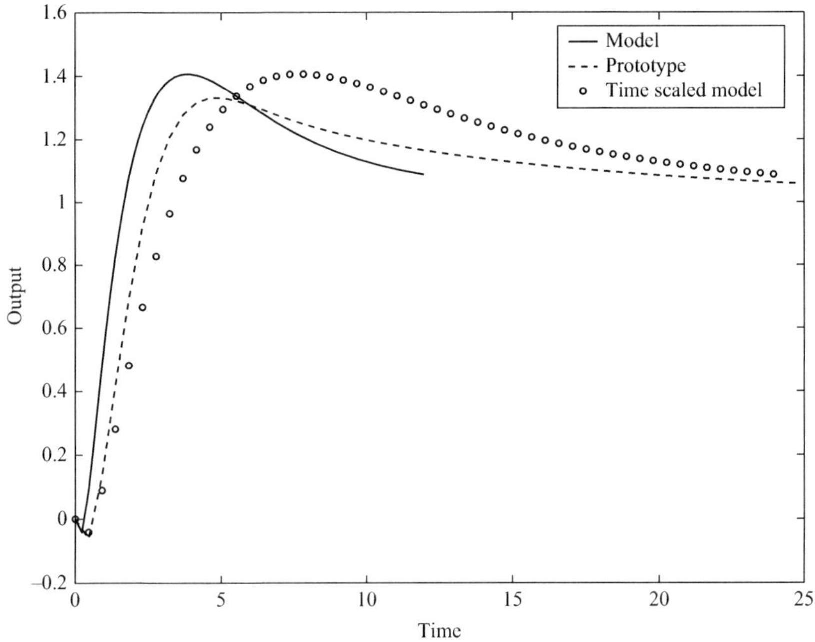

Figure 6.4 *Nonhomogeneous tuning rule step responses. When two distinct but dimensionally similar plants, the prototype $G_p(s)$ and the model $G_m(s)$, are controlled by proportional integral derivative (PID) control tuned with the Wang and Jin nonhomogeneous tuning rule, the closed loop behavior is no longer dimensionally similar. In fact at the sight of the closed loop step responses for the model (-) and the prototype (- -), the time scaled model response (o) is no longer equal to the prototype. Moreover, both overshoots are distinct, hence dimensional similarity is discarded*

case both responses are no longer similar because they cannot be made equal by time scaling. Note also that both step responses have distinct overshoot.

However, consider now a model and a prototype with the same time scaling factor $S_t = 2$ but with new gain scaling factor $S_k = 1$, thus both model and prototype have the same gain. Figure 6.5 shows the closed loop step response. Now, both closed loop responses show again the similarity property, because the nonhomogeneity was caused by the controller gain. However, note that for distinct gains similarity does no longer hold. As a conclusion nonhomogeneous tuning rules map dimensionally similar plants into nondimensionally similar closed loops, thus nonhomogeneous tuning rules do not preserve similarity. In this case prototype controller design based on controlled model experiments is infeasible. Moreover, the controller maintenance of families of dimensionally similar products that uses nonhomogeneous tuning rules is made unnecessarily complex.

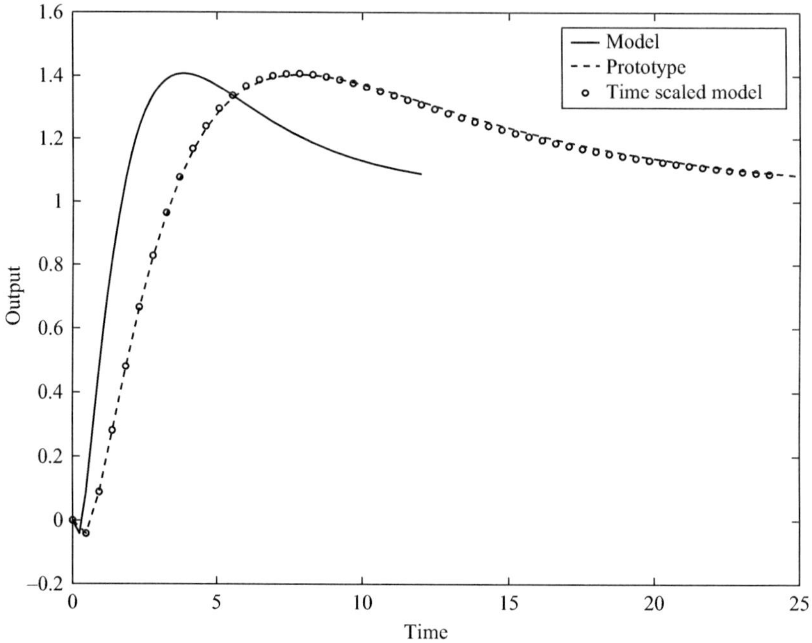

Figure 6.5 Nonhomogeneous tuning rule step responses. The nonhomogeneous Wang and Jin closed loop step responses for the model (-) and the prototype (- -) are in this case dimensionally similar as can be checked by the time scaled model (o). The reason is that Wang and Jin tuning rule is nonhomogeneous for the controller gain k_C but homogeneous for the integral τ_I and derivative time τ_D. In this case both model and prototype have the same gain value, so the gain scaling factor S_k is equal to unity. As a result the closed loops are dimensionally similar

References

[1] J.C. Doyle, B.A. Francis, and A.R. Tannenbaum. *Feedback Control Theory*. Macmillan Publishing Company, 1992

[2] A. O'Dwyer. *Handbook of PI and PID Controller Tuning Rules*. Imperial College Press, 2006

[3] G.E. Rotstein and D.E. Lewin. 'Simple PI and PID tuning for open-loop unstable systems.' *Industrial Engineering Chemistry Research*, 30:1864–1869, 1991

[4] S. Skogestad and I. Postlethwaite. *Multivariable Feedback Control. Analysis and Design*. Wiley, 1996

[5] T. Szirtes and P. Rózsa. *Applied Dimensional Analysis and Modeling*. Butterworth-Heinemann, 2007

[6] H. Wang and X. Jin. 'Direct synthesis approach of PID controller for second-order delayed unstable processes.' In *Proceedings of the 5th World Congress on Intelligent Control and Automation*, Hangzhou, China, pages 19–23, 2004
[7] J.G. Ziegler and N.B. Nichols. 'Optimum settings for automatic controllers.' *Transactions ASME*, 64, 1942

Chapter 7

Dimensionless PID tuning rules comparison

7.1 Introduction

The PID tuning is one of the most important issues in control theory and engineering due to the spread use of PID control in the industry [1]. In fact there are hundreds of tuning rules, derived in many different ways and with distinct characteristics [6].

In the same way, there are even more industrial processes each one having its own properties and particularities. Due to this variability it is well known that no single tuning rule is desirable in all instances [5]. On the contrary, there are tuning rules that are more suited for certain processes characteristics. As a result, the question that arises to the control engineer is to decide which tuning rule is better for the control engineering problem at hand. This topic of practical importance has been long recognized although the received attention in the bibliography has been scarce [4].

In this chapter we propose a framework for PID tuning rules comparison by using dimensional analysis. The proposal establishes the framework in which to perform the analysis for any user-selected indexes. The key point enabling the comparison framework yields in the dimensionless representation of both transfer function models and transfer function controllers. As a result it follows that the indices used for comparison are dimensionless with the following benefits:

- The set of dimensionless parameters is the minimum size set in which comparison can be performed exactly. Any other parameters set must be of greater size, thus complicating unnecessarily the comparison procedure.
- Each dimensionless parameter has a physical significance to be exploited in the comparison process. Moreover, dimensionless parameters avoid the problem of relative values of dimensional parameters. For instance, given an first order plus dead time (FOPDT) model with delay h equal to 1 second, is the delay a control issue? As stated, the question is meaningless because the answer depends on the value h/τ, being τ the time constant.
- The fact that comparison indices are dimensionless also implies that the indices no longer refer to a single model but to an infinite set of plants, the set of all dimensional similar plants. As a result the comparison results increase generality.
- Finally, the extrapolation of comparative indices to the particular problem at hand is straightforward, requiring a mere index scaling by a constant factor.

7.2 Elements of the comparative framework

Given the control structure of Figure 7.1, the plant $G(s)$, and a set of PID tuning rules, we want to assess and compare the performance of each tuning rule in the most meaningful and economic way. The scope of the comparison procedure considers:

- Plants $G(s)$: The set of linear time invariant (LTI) plants of arbitrary order. For instance in Table 7.1 we present some of the most common process plants used in control. This includes delayed second order plants minimum and non-minimum phases. The complexity of each plant, measured as the number of dimensionless numbers Π_G, is also presented in Table 7.1. Note that for

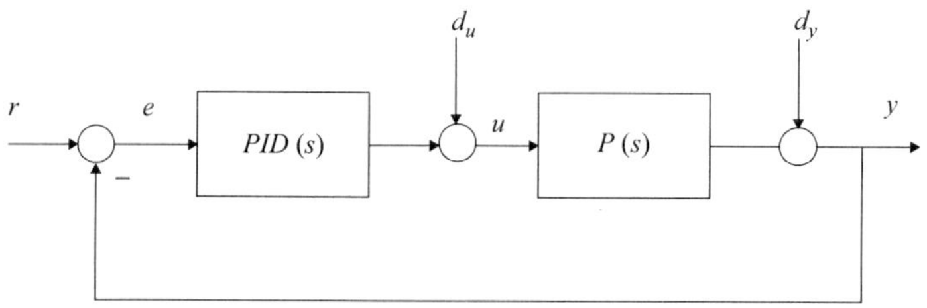

Figure 7.1 Control structure

Table 7.1 Plants considered

Dimension	Abbreviation	Plant	Dimensionless numbers
$\Pi_G \in R^1$	FOPDT	$\frac{k}{(\tau s+1)}e^{-hs}$	h/τ
	SO	$\frac{k}{(\tau_1 s+1)(\tau_2 s+1)}$	τ_2/τ_1
$\Pi_G \in R^2$	SOPDT	$\frac{k}{(\tau_1 s+1)(\tau_2 s+1)}e^{-hs}$	h/τ_1, τ_2/τ_1
	SOPIR	$\frac{k(-\tau_3 s+1)}{(\tau_1 s+1)(\tau_2 s+1)}$	τ_2/τ_1, τ_3/τ_1
	TO	$\frac{k}{(\tau_1 s+1)(\tau_2 s+1)(\tau_3 s+1)}$	τ_2/τ_1, τ_3/τ_1
$\Pi_G \in R^3$	SOPIRDT	$\frac{k(-\tau_3 s+1)}{(\tau_1 s+1)(\tau_2 s+1)}e^{-hs}$	h/τ_1, τ_2/τ_1, τ_3/τ_1
	TOPDT	$\frac{k}{(\tau_1 s+1)(\tau_2 s+1)(\tau_3 s+1)}e^{-hs}$	h/τ_1, τ_2/τ_1, τ_3/τ_1
	TOPIR	$\frac{k(-\tau_4 s+1)}{(\tau_1 s+1)(\tau_2 s+1)(\tau_3 s+1)}$	τ_2/τ_1, τ_3/τ_1, τ_4/τ_1
	FO	$\frac{k}{(\tau_1 s+1)(\tau_2 s+1)(\tau_3 s+1)(\tau_4 s+1)}$	τ_2/τ_1, τ_3/τ_1, τ_4/τ_1

comparative purposes, it is equally complex to compare an FOPDT plant with an second order (SO) plant.

- Controller tuning rules: Any controller tuning rule that can be expressed in dimensionless form, that is any homogeneous tuning rule. Although there are nonhomogeneous tuning rules, from the results of previous chapter and the literature review [6], the vast majority of existing tuning rules are homogeneous.
- Comparative indices: Any index derived from the open loop transfer function $C(s)G(s)$. This is very general because it includes the following indexes as particular instances:
  - $IAE = \int_0^\infty |e(t)|dt$
  - $ISE = \int_0^\infty e(t)^2 dt$
  - $TV = \int_0^\infty |\dot{u}(t)|dt$
  - Sensitivity function bandwidth w_B, $|S(jw_B)| = 1/\sqrt{2}$.
  - Sensitivity function peak $M_s = \max_w |S(jw)|$.

7.3 Dimensionless comparative framework

The dimensionless framework in which to perform the PID tuning rules comparison is presented. The comparison procedure is as follows:

1. For a given model $G(s)$ obtain the dimensionless transfer function $\bar{G}(\bar{s})$. Check the values of the dimensionless numbers. Look for large or small dimensionless numbers in order to assess their relative influence and possibly discard them (e.g. relative fast poles).
2. For a given tuning rule obtain its dimensionless form by calculating the functions Φ_1, Φ_2, and Φ_3, as follows:

$$k_C k \tau_{p1}^l = \Phi_1(\Pi_G, \Pi_\lambda) \tag{7.1}$$

$$\frac{\tau_I}{\tau_{p1}} = \Phi_2(\Pi_G, \Pi_\lambda) \tag{7.2}$$

$$\frac{\tau_D}{\tau_{p1}} = \Phi_3(\Pi_G, \Pi_\lambda) \tag{7.3}$$

   with τ_{p1} the time constant chosen for obtaining the dimensionless transfer function $\bar{G}(\bar{s})$.
3. Obtain the dimensionless loop transfer function $\bar{G}(\bar{s})\bar{C}(\bar{s})$.
4. Choose the indexes derived from the loop transfer function $\bar{G}(\bar{s})\bar{C}(\bar{s})$. Note that well-known indices as discussed above are obtained from $\bar{G}(\bar{s})\bar{C}(\bar{s})$.
5. Calculate and plot the indexes mentioned above as a function of the plant dimensionless numbers and the tuning parameters, if any. The set of dimensionless number is the minimum possible set.
6. Compare and select the tuning rules by means of the previously calculated indexes, using additional information on particular values of dimensionless numbers.

7. Recall that the information is in dimensionless form. However, dimensional values for indices comparison are extracted by mere scaling in order to obtain their dimensional value.

7.4 Dimensionless elements

In this section we review and derive the form of the dimensionless elements required for the comparison framework.

7.4.1 Loop transfer function

As shown in Chapter 6, the open loop transfer function $G(s)C(s)$ is a function of the plant dimensionless numbers Π_G and of the specifications dimensionless numbers Π_λ, if any, thus

$$G(s)C(s) = L(\bar{s}, \Pi_G, \Pi_\lambda) \tag{7.4}$$

7.4.2 Dimensionless closed loop transfer functions

Recall also from Chapter 6 that the relation between dimensional and dimensionless closed loop transfer function is

- $S(s) = S(\bar{s}, \Pi_G, \Pi_C)$
- $T(s) = T(\bar{s}, \Pi_G, \Pi_C)$
- $S_C(s) = k_C\bar{S}_C(\bar{s}, \Pi_G, \Pi_C)$
- $S_G(s) = k\tau_{p1}^{l}\bar{S}_G(\bar{s}, \Pi_G, \Pi_C)$

7.4.3 Dimensionless integral errors

The relation between integrals of dimensional and dimensionless functions is given by the following lemma:

Lemma 7.1. *Given the dimensional function $f(t)$ with dimensionless form given by $S_k\bar{f}(\bar{t})$ with $t = S_t\bar{t}$, and S_k, S_t constants, then the integral*

$$\int_{t_0}^{t_f} f(t)dt = S_kS_t \int_{\bar{t}_0}^{\bar{t}_f} \bar{f}(\bar{t})d\bar{t} \tag{7.5}$$

with integration limits $\bar{t}_0 = t_0/S_t$ and $\bar{t}_f = t_f/S_t$.

Proof. The result follows by direct substitution of $f(t)$ by $S_k\bar{f}(\bar{t})$ and t by $S_t\bar{t}$. □

This general result is particularized to integral errors and to integral control actions, which are well-known performance measures of controlled loops.

Integral errors

Given the closed loop error $e(t) = r(t) - y(t)$, a classical performance measure is given by the integration of distinct error functions $f(e(t))$, that is

$$\int_{t_0}^{t_f} f(e(t))dt \tag{7.6}$$

Depending on the function $f(.)$ we have distinct integral errors such as

- IAE when $f(e(t)) = |e(t)|$
- ISE when $f(e(t)) = e(t)^2$

Lemma 7.2. *Given the dimensional error signal due to output disturbance, i.e. $e(t) = S(t)d(t)$, with $d(t)$ the disturbance signal at the output, the value of the integral error is given by*

$$\int_{t_0}^{t_f} f(e(t))dt = k_d S_t^{c+1} \int_{\bar{t}_0}^{\bar{t}_f} f(\bar{e}(\bar{t}))d\bar{t} \tag{7.7}$$

with k_d the disturbance gain, c the number of integrators in $S(s)d(s)$, and $\bar{e}(\bar{t})$ the dimensionless error $\bar{e}(\bar{t}) = \mathcal{L}^{-1}\{\bar{S}(\bar{s})\bar{d}(\bar{s})\}$.

The result provides, for a whole set of similar plants and similar disturbances, the values of the dimensionless integral error as a function of the transfer function dimensionless numbers $\bar{e}(\bar{t})$. In order to obtain the dimensional value of the error $e(t)$ it is simply required to multiply $\bar{e}(\bar{t})$ by $k_d S_t^{c+1}$, which only depends on the disturbance properties.

Integral control actions

Another interesting integral refers to the variations of control actions $TV = \int_0^\infty |\dot{u}(t)|dt$, as it is a measure of the control activity level.

Lemma 7.3. *Given the dimensional control action signal (i.e. $u(t) = S_C(t)d(t)$), the relation between dimensional and dimensionless integrals is*

$$\int_{t_0}^{t_f} |\dot{u}(t)|dt = k_d k_C S_t^{c-1} \int_{\bar{t}_0}^{\bar{t}_f} |\dot{\bar{u}}(\bar{t})|d\bar{t} \tag{7.8}$$

where k_C is the controller gain, k_d the disturbance gain, and c the number of integrators in $S_C(s)d(s)$.

Proof. The proof follows by noting that $u(t) = S_C(s)d(s)$ and the dimensionless form is given by $\bar{u}(\bar{t}) = k_d k_C S_t^c \bar{S}_C(\bar{s})\bar{d}(\bar{s})$. Finally consider the change of variable $t = S_t\bar{t}$ in the integral and the result follows. □

The only difference with the preceding case is that now the dimensional form depends also on the controller gain k_C.

7.4.4 Indexes

In this section the dimensionless indexes to perform the comparison are presented. Despite the variety of possible indexes an important point is that the set of indexes used should capture the fundamental trade-off of any control design. The dimensionless indexes considered in the following are:

- Dimensionless crossover frequency $\bar{w}_c$. $|\bar{G}\bar{C}(j\bar{w}_c)| = 1$.
- Dimensionless closed loop bandwidth $\bar{w}_B$. $|\bar{S}(j\bar{w}_B)| = 1/\sqrt{2}$. It is an indicator of disturbance rejection performance.
- Dimensionless complementary sensitivity bandwidth $\bar{w}_{BT}$. $|\bar{T}(j\bar{w}_{BT})| = 1/\sqrt{2}$. It is an indicator of tracking performance and noise mitigation.
- Peak of dimensionless sensitivity function $\bar{M}$. $\bar{M} = \max_{\bar{w}} |\bar{S}(j\bar{w})|$. It is an indicator of closed loop robustness toward model uncertainty.
- Dimensionless integral absolute error $\int_{\bar{t}_0}^{\bar{t}_f} |\bar{e}(\bar{t})d\bar{t}|$.
- Dimensional variation of the control input $k_C \int_{\bar{t}_0}^{\bar{t}_f} |\dot{\bar{u}}(\bar{t})|d\bar{t}$.

However, it should be remarked that any index derived from the dimensionless loop transfer function just depends on the plant and controller dimensionless numbers.

7.5 Application example

In this section we compare tuning rules for FOPDT models that require a performance tuning parameter, the desired closed loop time constant λ. The tuning rules considered are:

- *Skogestad*. Analytic tuning rules are derived from internal model control (IMC) considerations in Reference 8. In fact the derived controller for FOPDT models is a proportional integral (PI). The objective is to obtain simple but still good performing tuning. The tuning rules are
  - $k_C = \tau/k(\lambda + h)$
  - $\tau_I = \min\{h, 4(\lambda + h)\}$
  - $\tau_D = 0$
- *Rivera et al.* The PID tuning rule proposed in Reference 7 is based on the IMC principle assuming an FOPDT model with the delay term approximated by a first order Padé. The tuning rules are
  - $k_C = (\tau + 0.5h)/k(\lambda + 0.5h)$
  - $\tau_I = \tau + 0.5h$
  - $\tau_D = \tau h/(2\tau + h)$
- *Dahlin*. The tuning rules of Reference 2 are equal to the ones proposed by Rivera *et al.* except for the k_C term:
  - $k_C = (\tau + 0.5h)/k(\lambda + h)$
  - $\tau_I = \tau + 0.5h$
  - $\tau_D = \tau h/(2\tau + h)$

- *Lee et al.* In Reference 3 the tuning rules proposed are also derived from the IMC principles.
  - $k_C = \tau_I/k(\lambda + h)$
  - $\tau_I = \tau + h^2/2(\lambda + h)$
  - $\tau_D = h^2(1 - \frac{h}{3\tau_I})/2(\lambda + h)$

7.5.1 PID tuning rules dimensionless characterization

In this section we characterize the above mentioned tuning rules by means of dimensionless parameters $\bar{h} = h/\tau$ and $\bar{\lambda} = \lambda/\tau$. Note also that $l = 0$, then

$$
\begin{aligned}
k_C k &= \Phi_1\left(\frac{h}{\tau}, \frac{\lambda}{\tau}\right) \\
\frac{\tau_I}{\tau} &= \Phi_2\left(\frac{h}{\tau}, \frac{\lambda}{\tau}\right) \\
\frac{\tau_D}{\tau} &= \Phi_3\left(\frac{h}{\tau}, \frac{\lambda}{\tau}\right)
\end{aligned}
$$

In Table 7.2 we present the values $k_C k$, τ_I/τ, and τ_D/τ for each one of the tuning rules.

In what follows we perform the comparison of the dimensionless indices considered.

7.5.2 Dimensionless sensitivity bandwidth comparison

In this section we compare the bandwidth provided by each one of the tuning rules as a function of the controllability index and the control relaxation. The results can be seen in Figure 7.2 and the following conclusions can be established:

- Skogestad, Rivera, and Dahlin provide very similar bandwidth values, despite Skogestad proposes a PI and Rivera and Dahlin a PID.
- Lee *et al.* behave in a similar way as Skogestad, Rivera, and Dahlin but the bandwidth values provided are lower.

In general there are neither qualitative nor quantitative differences among the sensitivity function bandwidth provided by the four tuning rules.

Table 7.2 Dimensionless form of the PID tuning rules

Tuning rule	$\mathbf{\Phi_1 = kk_C}$	$\mathbf{\Phi_2 = \tau_I/\tau}$	$\mathbf{\Phi_3 = \tau_D/\tau}$
Skogestad	$1/(\bar{\lambda} + \bar{h})$	$\min[1, 4(\bar{\lambda} + \bar{h})]$	0
Rivera	$(1 + 0.5\bar{h})/(\bar{\lambda} + 0.5\bar{h})$	$1 + 0.5\bar{h}$	$\bar{h}/(2 + \bar{h})$
Dahlin	$(1 + 0.5\bar{h})/(\bar{\lambda} + \bar{h})$	$1 + 0.5\bar{h}$	$\bar{h}/(2 + \bar{h})$
Lee *et al.*	$(\bar{\lambda} + (3/2)\bar{h})/(\bar{\lambda} + \bar{h})^2$	$1 + (\bar{h})^2/(2(\bar{\lambda} + \bar{h}))$	$\dfrac{\bar{h}(3 + 4(\bar{\lambda} + \bar{h}))}{((\bar{\lambda} + \bar{h})(6(\bar{\lambda} + \bar{h}) + 2\bar{h}))}$

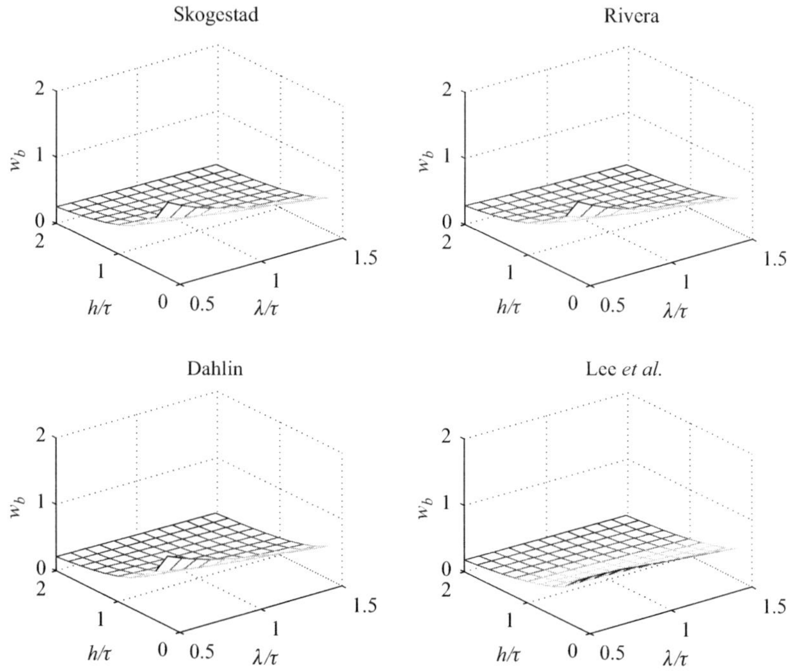

Figure 7.2 Sensitivity function bandwidth w_b

7.5.3 Dimensionless sensitivity peak comparison

Figure 7.3 shows the sensitivity peak function of each tuning rule as a function of the controllability index and the control relaxation. It can be seen that:

- Skogestad and Dahlin have a very similar behavior of the sensitivity function peak.
- Rivera has a similar behavior for lower $\bar{h}$ values than the other tuning rules but as $\bar{h}$ increases Rivera rapidly increases the sensitivity peak.
- Lee *et al.* provide in general the lowest sensitivity peak. However, for small $\bar{\lambda}$ and $\bar{h}$ it provides higher sensitivity peaks than the other tuning rules.

In summary, each tuning rule provides a high sensitivity function peak at distinct points.

7.5.4 Dimensionless integral absolute error

Figure 7.4 shows the IAE for the tuning rules. The points of interest are:

- Dahlin provides in general the lower IAE for all $\bar{\lambda}$ and $\bar{h}$.

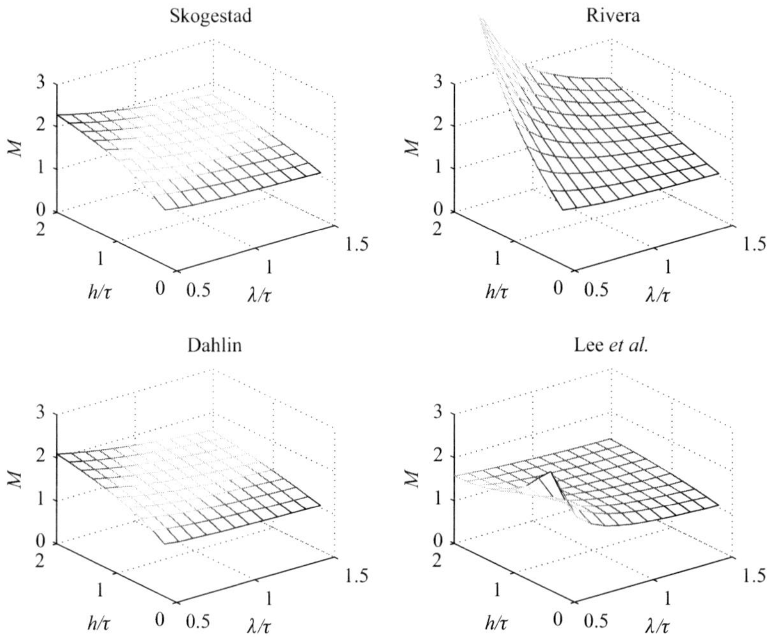

Figure 7.3 Sensitivity function peak M

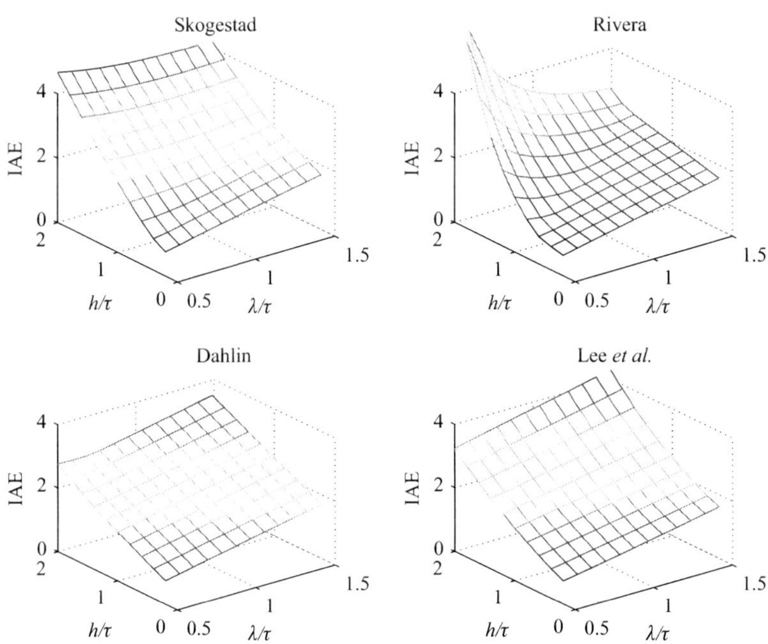

Figure 7.4 Integral absolute error

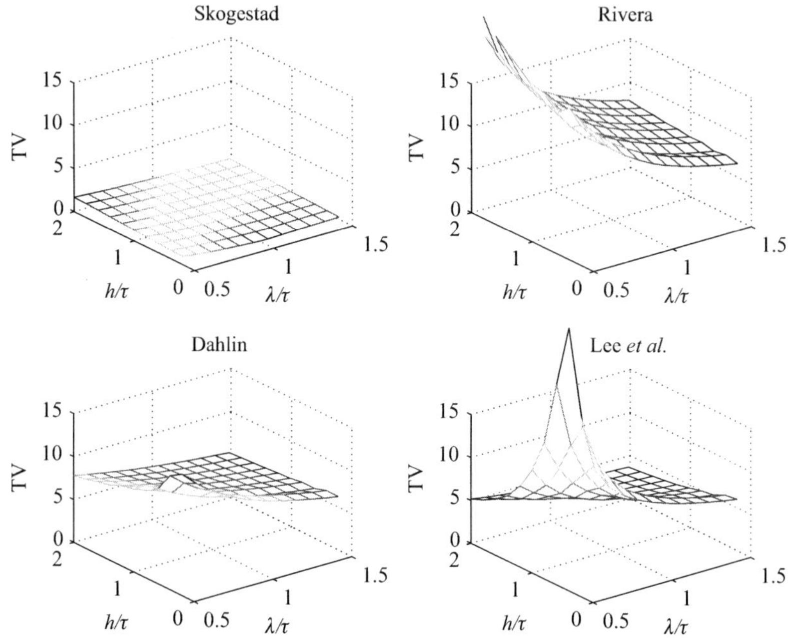

Figure 7.5 Control action variability

- Rivera and Lee *et al.* provide the highest peaks of IAE. Rivera peaks for high $\bar{h}$ and low $\bar{\lambda}$ while Lee *et al.* peaks for high $\bar{h}$ and high $\bar{\lambda}$.
- Skogestad IAE increases almost linearly with $\bar{h}$.

7.5.5 Dimensionless control action variation

Regarding the control action variability presented in Figure 7.5 we can establish that:

- The lowest TV corresponds to Skogestad. The other tuning rules provide higher values.
- Dahlin tuning rule provides a nearly constant value of TV for all $\bar{h}$ and $\bar{\lambda}$.
- Rivera provides higher values of TV for higher $\bar{h}$ and lower $\bar{\lambda}$ while Lee *et al.* provide high TV values for lower $\bar{h}$ and $\bar{\lambda}$.

7.6 PID tuning rules selection

In this section we present by means of an example the practical utility of the proposed comparative framework in order to select the most appropriate tuning rule to the control problem particularities.

Given the following FOPDT model:

$$G(s) = \frac{1}{5s + 1} e^{-2s} \tag{7.9}$$

the controllability index is $h/\tau = 2/5 = 0.4$. In Figure 7.6 the four indices considered are represented, that is the bandwidth w_b, the sensitivity peak M, the absolute integral error IAE, and the control action variability TV for each one of the tuning rules considered.

At the sight of Figure 7.6 we can study two scenarios: first tunings with $\lambda/\tau < 1$, that is closed loops faster than the open loop dynamics. In this scenario, for the indices considered, the most suited choice is the Skogestad tuning rule because it provides a very low control action variability TV while keeping the rest of indices very similar to Rivera and Dahlin tuning rules. In the second scenario with $\lambda/\tau > 1$, that is closed loop slower than open loop dynamics, Skogestad still provides the minimum control variability value TV; however, Lee *et al.* are able to provide a lower IAE. As a result, depending on the considered cost function the choice would be between Skogestad and Lee *et al.*

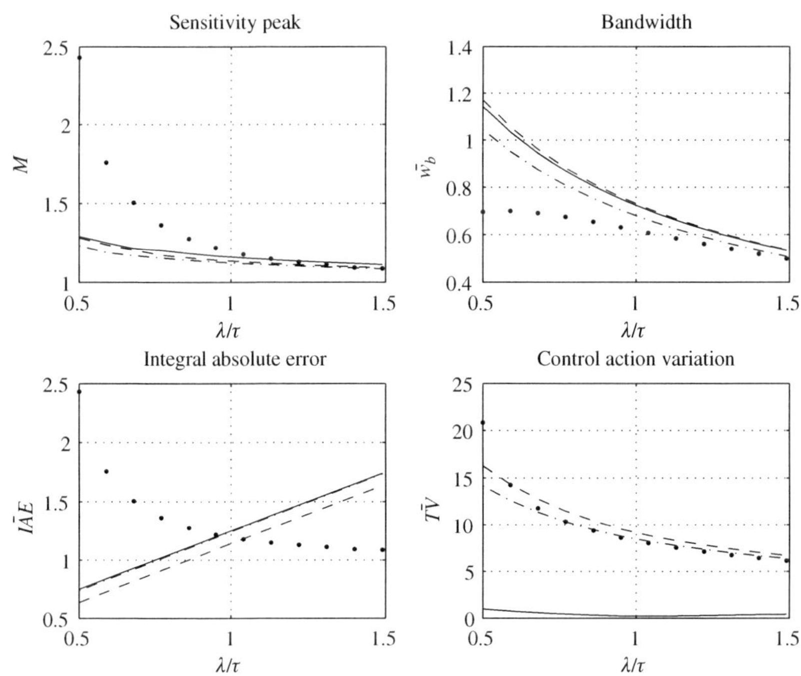

Figure 7.6 Dimensionless indices for '-' Skogestad, '- -' Rivera, '- ·' Dahlin, and '·' Lee et al.

References

[1] K.J. Åström and T. Hägglund. *PID Controllers: Theory, Design and Tuning.* Instrument Society of America, 1995

[2] E.G. Dahlin. 'Designing and tuning digital controllers.' *Instrumentaion and Control Systems*, 41(6):77–81, 1968

[3] Y. Lee, M. Lee, S. Park, and C. Brosilow. 'PID controller tuning for desired closed-loop responses for SISO systems.' *AIChE*, 44(1):106–115, 1998

[4] A. Leva, C. Cox, and A.E. Ruano, 'Hands-on PID autotuning: a guide to better utilisation,' ed. 1ª, Austria: IFAC, 2002

[5] G.K. McMillan. *Good Tuning: A Pocket Guide*. The Instrumentation, Systems And Automation Society, 2000

[6] A. O'Dwyer. *Handbook of PI and PID Controller Tuning Rules.* Imperial College Press, 2006

[7] D. Rivera, M. Morari, and S. Skogestad. 'Internal model control: 4. PID controller design.' *Industrial Engineering Process Design and Developments*, 25:252–265, 1986

[8] S. Skogestad. 'Simple analytic rules for model reduction and PID controller tuning.' *Journal of Process Control*, 13(4):291–309, 2003

Chapter 8
Control of dimensionally similar systems

with A. Claramonte

8.1 Introduction

The use of scaled models is prevalent in engineering applications because they allow for experimental tests to be performed without requiring the physical system (i.e. prototype) to be constructed. The model is built dimensionally similar with respect to the prototype. Consequently, the experimental results on the model can be extrapolated to the prototype for analysis and design purposes [2,11]. Dimensional analysis applications can be found in fluid mechanics [12], chemistry [13], and robotics. For instance, dimensionally similar models have been used in robotics for optimum design of jumping robots [8] and for study of energy efficiency of walking machines [9].

The analysis, evaluation, and design of control systems can also be benefitted by using dimensionally similar models. The use of models permits for experimental tests to be performed when the prototype does not exist or it is difficult or dangerous to perform experiments on it. Moreover, there are families of manufactured products that, by construction, share some similarity relation. From a control perspective the objective is to find an expedient way for analysis, design, and maintenance of controllers for the whole family of products.

In the bibliography we may find particular applications of dimensional analysis to control design on vehicle dynamics [4] and robotics [7]. In Reference 4 a scaled testbed is constructed for vehicle control application, showing the dynamic similarity with actual vehicles. The design of a robust controller for distinct vehicles is investigated in Reference 5 from a dimensionless approach. Another contribution is presented in Reference 7 where scaling laws for continuous time and discrete time controllers are developed for a single flexible link manipulator. The approach taken in all cases considers dimensionless numbers derived from physical parameters. However, it is difficult to generalize this approach because on the one hand the dimensionless numbers depend on physical parameters such as inertias or capacitances, and, on the other hand, the number of parameters forming the pi groups is problem dependent.

However, the case is different for the transfer function framework. In the transfer function approach physical parameters are transformed into gains, time delays, poles, and zeroes time constants [1]. In this way, only two scaling factors are required to relate two dimensionally similar transfer functions, the gain scaling factor S_k, and the time scaling factor S_t. Although we always have to consider physical parameters when applying dimensional analysis and similarity, once dimensional similarity is established, the transfer function framework greatly simplifies and generalizes the scaling of controllers.

The analysis and design of control systems in the transfer function framework are pervasive in control applications because normally the system state is not fully measurable and dynamical controllers are required. In this case, for continuous time dimensionally similar systems controlled by means of digital controllers, a similar performing control level can be achieved by modifying the controller in two straightforward ways: (i) by adjusting the controller gain and (ii) by adjusting the controller sampling time. Moreover, the adjusting factors are directly given by the system scaling factors.

However, the construction of fully similar models is costly and sometimes it is not even possible [12]. In this case we no longer have complete similarity but partial similarity. In this chapter we show how partial similarity can be treated using robust control theory by modeling the lack of similarity as uncertainty.

8.2 Control of dimensionally similar systems

Consider a model defined mathematically by the transfer function $G_m(s_m)$, which is dimensionally similar to a prototype $G_p(s_p)$. Assume that the controller $C_m(s_m)$ has been designed and tested on the model and that it achieves a desired closed loop performance. The point to be investigated is, based on the knowledge of $C_m(s_m)$, the determination of the controller $C_p(s_p)$ applied to the prototype (see Figure 8.1) that achieves a similar closed loop performance.

The second problem to be considered is the assessment of the resulting prototype controller robustness properties when the dimensional similarity is partial. In

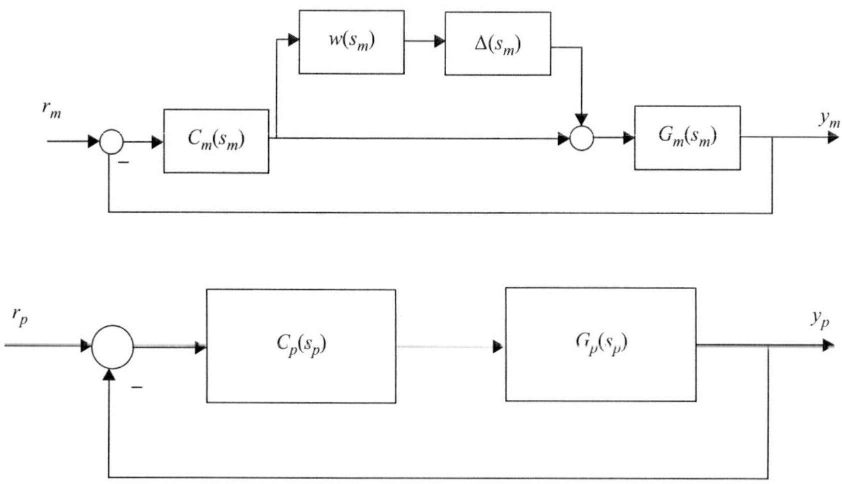

Figure 8.1 Model and prototype control loops

fact there are instances that it is very difficult and even impossible to design a model that is completely similar to a prototype. The best that can be achieved is partial similarity. In this case we may consider that the model is uncertain, represented as multiplicative uncertainty in Figure 8.1.

8.3 Complete similarity

In this section we consider the problem of controller design for systems that follow a similarity relation, when complete dimensional similarity exists. If we design a controller in order to fulfill some performance level on a model, then the question is how the currently existing model controller should be modified in order to achieve a similar performance level on the prototype system.

8.3.1 Continuous time control

The model-prototype controllers equivalence is presented by the following theorem:

Theorem 8.1. *Consider two continuous time transfer functions $G_p(s_p)$ and $G_m(s_m)$. Assume that $C_m(s_m)$ is the continuous time model controller achieving the desired transfer function $L_m(s_m) = G_m(s_m)C_m(s_m)$. If the prototype controller $C_p(s_p)$ is obtained as $C_p(s_p) = S_k^{-1}S_t^{-l}C_m(S_ts_p)$ then, if $G_p(s_p)$ and $G_m(s_m)$ are continuous time similar, then the open loop transfer functions of the model and the prototype are similar, that is $L_m(\bar{s}) = L_p(\bar{s})$.*

Proof. In case of continuous time similarity we have

$$\begin{aligned}
L_p(s_p) &= G_p(s_p)C_p(s_p) \\
&= S_kS_t^lG_m(S_ts_p)S_k^{-1}S_t^{-l}C_m(S_ts_p) \\
&= G_m(S_ts_p)C_m(S_ts_p) \\
&= L_m(S_ts_p) \\
&= L_m(s_m)
\end{aligned} \tag{8.1}$$

This implies that $L_p(\bar{s}) = L_m(\bar{s})$, that is model and prototype open loop transfer functions are equal except for a time scaling factor. □

In words Theorem 8.1 states that the only modifications to be performed to an existing model controller are to modify its gain with $S_k^{-1}S_t^{-l}$ and to change the model controller Laplace variable by S_ts_p.

8.3.2 Discrete time control

The model-prototype controllers equivalence on sampled data control is presented by the following theorem:

Theorem 8.2. *Consider two sampled data transfer functions $G_p(z_{SD})$ and $G_m(z_{SD})$, with sampling times $(T_s)_p$ and $(T_s)_m$, respectively. Assume that $C_m(z_{SD})$ is the discrete time model controller with sampling time $(T_s)_m$ achieving the desired transfer function $L_m(z_{SD}) = G_m(z_{SD})C_m(z_{SD})$. Now if the prototype controller $C_p(z_{SD})$ is obtained as $C_p(z_{SD}) = S_k^{-1}S_t^{-l}C_m(z_{SD})$ with sampling time $(T_s)_p = S_t(T_s)_m$, then if $G_p(z_{SD})$ and $G_m(z_{SD})$ are sampled-data similar then $L_p(z_{SD}) = L_m(z_{SD})$.*

Proof. In the case of discrete time similarity we have $G_p(z) = S_kS_t^lG_m(z)$, so the loop transfer function $L_p(z)$ can be written as

$$\begin{aligned} L_p(z) &= G_p(z)C_p(z) \\ &= S_kS_t^lG_m(z)S_k^{-1}S_t^{-l}C_m(z) \\ &= G_m(z)C_m(z) \\ &= L_m(z) \end{aligned} \tag{8.2}$$

Then the discrete transfer functions $L_m(z)$ and $L_p(z)$ are equal in a strict discrete time sense. Moreover, due to sampled data similarity it holds that the dimensionless number z_{SD} is also equal in the model and the prototype, that is $z_{SD} = e^{s_p(T_s)_p} = e^{s_m(T_s)_m}$. □

Theorem 8.2 states that the only modifications to be performed to an existing controller are to modify its gain by means of the inverse gain scaling factor $S_k^{-1}S_t^{-l}$ and to scale the prototype controller sampling time by the time scaling factor S_t.

Application example

Consider two continuous transfer functions, the model and the prototype, given by

$$G_m(s) = \frac{(15s+1)(0.005s+1)}{(3.5s+1)(0.5s+1)(0.1s+1)}$$

$$G_p(s) = \frac{5(24s+1)(0.008s+1)}{(5.6s+1)(0.8s+1)(0.16s+1)}$$

We have five time constants (three poles and two zeros) and the dimensionless numbers are τ_{p2}/τ_{p1}, τ_{p3}/τ_{p1}, τ_{z1}/τ_{p1}, τ_{z2}/τ_{p1}, and $\tau_{p1}s$.

In Table 8.1 the dimensionless numbers for the model and the prototype are shown. In can be seen that τ_{p2}/τ_{p1}, τ_{p3}/τ_{p1}, τ_{z1}/τ_{p1}, and τ_{z2}/τ_{p1} are equal, so we have complete similarity. The time scaling factor is given by $S_t = (\tau_{p1})_p/(\tau_{p1})_m = 5.6/3.5 = 1.6$. The gain scaling factor is given by $S_k = k_p/k_m = 5$.

Next, we calculate the zero order hold equivalent discrete transfer function of the model $G_m(z)$ and of the prototype $G_p(z)$. The model sampling time selected for the model is $(T_s)_m = 0.02$ s. The prototype sampling time is calculated as $(T_s)_p = S_t(T_s)_m = 1.6 \cdot 0.02 = 0.032$ s in order to obtain similarity in the discrete

Table 8.1 Dimensionless numbers of model and prototype

Dimensionless number	Model	Prototype
τ_{p2}/τ_{p1}	$0.5/3.5 = 0.14$	$0.8/5.6 = 0.14$
τ_{p3}/τ_{p1}	$0.1/3.5 = 0.03$	$0.16/5.6 = 0.03$
τ_{z1}/τ_{p1}	$15/3.5 = 4.28$	$24/5.6 = 4.28$
τ_{z2}/τ_{p1}	$0.005/3.5 = 0.001$	$0.008/5.6 = 0.001$
$\tau_{p1}s$	$3.5s_m$	$5.6s_p$

time transfer functions. The resulting discrete time transfer function for the model $G_m(z)$ is given by

$$G_m(z) = S_k^{-1}G_p(z) = \frac{0.02341z^2 - 0.01639z - 0.006977}{z^3 - 2.774z^2 + 2.556z - 0.7821} \tag{8.3}$$

On the basis of the above discrete time model we design a minimum time controller for the model $C_m(z)$, given by

$$C_m(z) = \frac{2.469 \cdot 10^4 z^3 - 6.849 \cdot 10^4 z^2 + 6.311 \cdot 10^4 z - 1.931 \cdot 10^4}{z^3 - 578z^2 + 404.7z + 172.3}, \quad (T_s)_m = 0.02 \tag{8.4}$$

The controller for the discrete time prototype is given by Theorem 8.2 resulting in

$$C_p(z) = \frac{4938z^3 - 1.37 \cdot 10^4 z^2 + 1.262 \cdot 10^4 z - 3863}{z^3 - 578z^2 + 404.7z + 172.3}, \quad (T_s)_p = 0.032 \tag{8.5}$$

In Figure 8.2 the outputs of the model and the prototype for a step input in the reference can be seen. In both cases it takes three samples to make the output to follow the reference, as expected. We can conclude that $C_p(z)$ is also minimum time controller for the prototype $G_p(z)$.

8.4 Partial similarity

Under partial similarity, model and prototype have distinct dimensionless representation (i.e. $\bar{G}_m(\bar{s}) \neq \bar{G}_p(\bar{s})$) because some dimensionless numbers have distinct value in the model and the prototype. For the sake of clarity we assume that model and prototype gains are known. In case both gains are not perfectly known the same approach can be used but adding an uncertain gain scaling factor. Moreover, time delays on the model and prototype are converted to rational expressions by Padé approximation.

The uncertainty is characterized as multiplicative dynamic uncertainty [10] because normally the dimensionless numbers related with dominant time constants

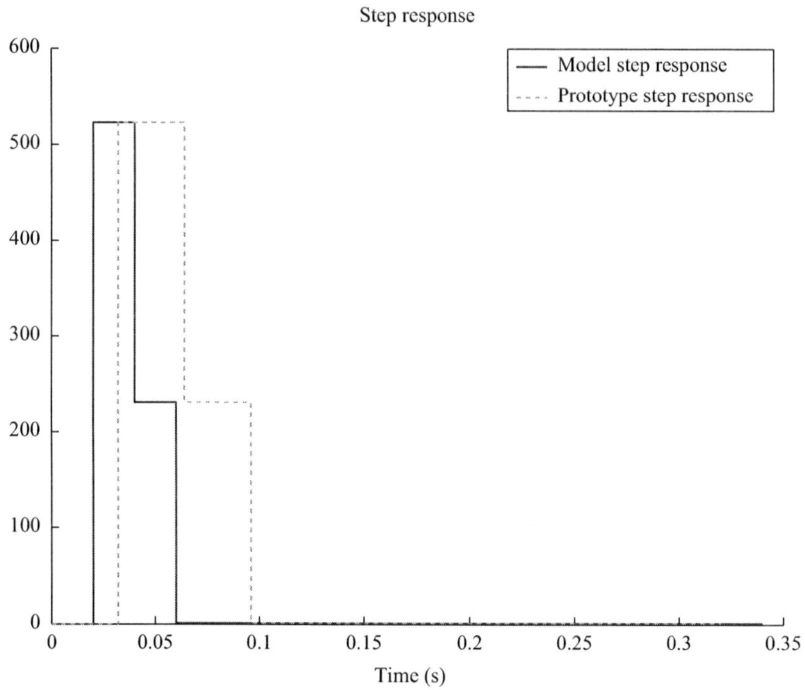

Figure 8.2 Model and prototype closed loop step responses

are similar, whereas dimensionless time constants related to fast dynamics are no longer similar. Moreover, dynamic uncertainty considers high-frequency unmodeled dynamics and it is computationally more tractable than the parametric counterpart [3]. Henceforth the uncertain model is

$$\bar{G}_m(\bar{s}) = \bar{G}_{m0}(\bar{s})(1 + W(\bar{s})\Delta(\bar{s})) \tag{8.6}$$

where the weight $W(\bar{s})$ is in general a rational transfer function and $\Delta(\bar{s})$ is any stable transfer function, which at any frequency is not larger than 1 in magnitude (i.e. $|\Delta(j\bar{w})| \leq 1, \ \forall\bar{w}$). The weight $W(\bar{s})$ can be calculated as an upper bound of $l(\bar{w})$, that is $|W(j\bar{w})| \geq l(\bar{w}), \ \forall\bar{w}$, where $l(\bar{w})$ is defined as

$$l(\bar{w}) = \max \left| \frac{\bar{G}_p - \bar{G}_{m0}}{\bar{G}_{m0}} \right| \tag{8.7}$$

Consequently, we have the nominal model $G_{m0}(\bar{s})$ together with the uncertainty description $|W(j\bar{w})| \geq l(\bar{w}), \ \forall\bar{w}$, so robust analysis and design procedures [10] can be applied. Finally we show that when a controller achieves robust stability (RS) and robust performance (RP) on the uncertain dimensionless model, it also achieves RS and RP on the dimensional prototype.

Theorem 8.3. *If the controller designed on the model $C_{m0}(\bar{s})$ achieves RS or RP on $G_m(\bar{s}) = G_{m0}(\bar{s})(1 + W(\bar{s})\Delta(\bar{s}))$, then the controller $C_p(\bar{s}) = S_k^{-1}S_t^{-l}C_{m0}(\bar{s})$ achieves RS or RP on $G_p(\bar{s})$.*

Proof. Assume that $C_{m0}(\bar{s})$ achieves RS, then $\|W(\bar{s})T_m(\bar{s})\|_\infty < 1$. The complementary sensitivity function is $T_m(\bar{s}) = G_{m0}(\bar{s})C_{m0}(\bar{s})(1 + G_{m0}(\bar{s})C_{m0}(\bar{s}))^{-1}$. Now by multiplying and dividing by $S_kS_t^l$ and recalling that $G_p(\bar{s}) = S_kS_t^lG_{m0}(\bar{s})$ and $C_p(\bar{s}) = S_k^{-1}S_t^{-l}C_{m0}(\bar{s})$ we have $T_p(\bar{s}) = G_p(\bar{s})C_p(\bar{s})(1 + G_p(\bar{s})C_p(\bar{s}))^{-1}$. Finally the change of variable $\bar{s} = S_ts_p$, which is a time scaling: does not modify the magnitude then $\|W(s)T^p(s)\|_\infty < 1$. The RP proof follows a similar reasoning and it is omitted. □

8.5 Experimental case study

The 'ecp Industrial Plant Emulator' [6] is a three-shaft system where drive and load inertias can be adjusted (see Figure 8.3). The transfer function between the angular output velocity and the motor drive voltage input is given by

$$\frac{\dot{\theta}}{V} = \frac{F}{Js + 1} \tag{8.8}$$

where F is the dynamic friction and J is the total system inertia.

Figure 8.3 Physical model configuration

Figure 8.4 Physical prototype configuration

We consider two systems, the model shown in Figure 8.3 and the prototype shown in Figure 8.4. The model has two weights: the driver pulley (the small one) and the load pulley, which fix the model inertia J_m.

The prototype is physically configured by adding two extra weights to each pulley, so the drive pulley and the load pulley have now four weights each (see Figure 8.4). As a result the prototype increases the system inertia with respect to the model.

First we identify model and prototype inertia by a simple step response with step magnitude $u = 0.6$ V, yielding $J_m = 10.2$ s, and $J_p = 16.25$ s. On the other hand, model and prototype gains are obtained by closed loop identification with a proportional controller $k_C = 0.1$ yielding $F_m = 115.93$ rad/sV and $F_p = 112.64$ rad/sV. The reason is that low step voltages rapidly saturate the motor drive and a closed loop experiment permits to identify the gains more accurately. The result confirms that model and prototype gains hardly vary because they are related to the dynamic friction. As a result, the gain scaling factor is $S_k = F_p/F_m = 1.029$ and the time scaling factor is $S_t = J_p/J_m = 1.59$.

Once the model is identified, a PI controller is designed for the model with settling time equal to 1 s and damping ratio of 0.7, which yields

$$C_m(s) = 2.66\frac{1 + 0.26s}{s} \tag{8.9}$$

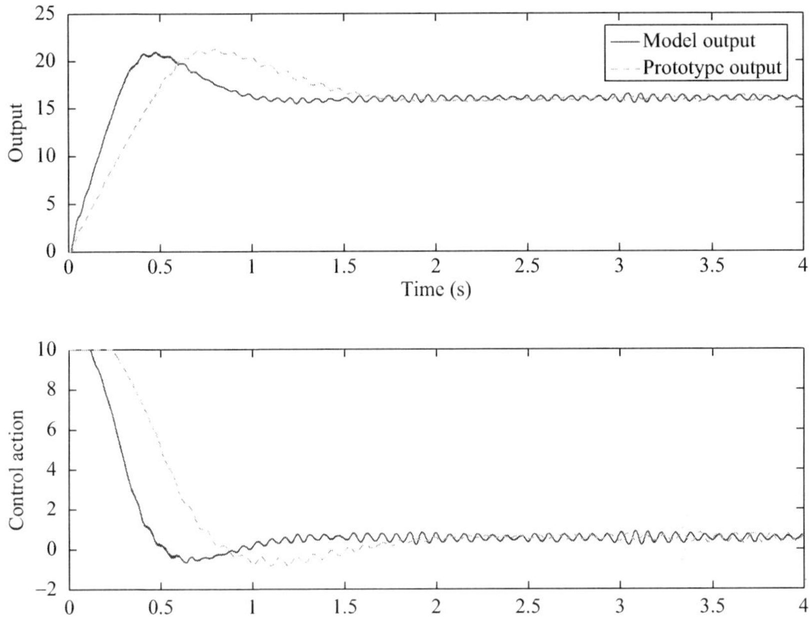

Figure 8.5 Model and prototype outputs and control actions

The controller is discretized by zero order hold, yielding

$$C_m(z) = \frac{0.6916z - 0.69}{z - 1} \tag{8.10}$$

The controller is digitally implemented on the computer, with sampling time of $T_S^m = 0.0006$ s. The model controlled output and model control action can be seen in Figure 8.5, achieving the design specifications.

Next we tackle the problem of prototype controller design. The prototype controller $C_p(s)$ is obtained by transformation of the model controller $C_m(s)$. Recall that if the discrete prototype controller is equal to $C_p(z_{SD}) = S_k^{-1} C_m(z_{SD})$ with $(T_s)_p = S_t(T_s)_m$, then model and prototype closed loops are similar. Because $S_K \approx 1$ and $S_t = 1.59$, we implement the same discrete digital controller (8.10) with the only modification being that the sampling time is set to $(T_s)_p = 0.00096$ s.

The prototype controlled output and prototype control action are shown in Figure 8.5. Now the prototype settling time is of 1.6 s, as expected. Moreover, the damping ratio is kept equal to 0.7 because the damping ratio is a dimensionless parameter, and thus equal under similarity.

We empirically show that model and prototype control outputs and control actions are similar. Recall that two outputs are similar if the only transformations required to make them equal are a gain scaling and a time scaling. Due to the fact

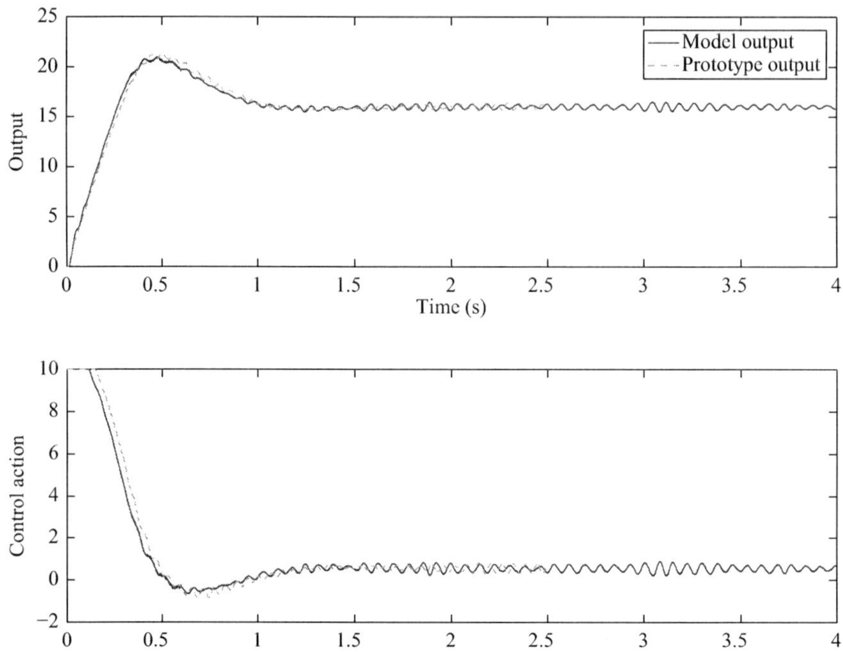

Figure 8.6 Prototype time scaled output

$S_k \approx 1$ just a time scaling by a factor of $S_t = 1.59$ is required. Figure 8.6 shows the model input and output together with the prototype input and output time scaled by $t_p = S_t^{-1} t_m$. The results show that in fact model and prototype closed loops are similar.

Finally we assume that there is partial similarity and consider that the prototype inertia is not perfectly known, that is $J_m = (1 \pm 10\%)J_p$. The dimensionless nominal model and the dimensionless prototype are

$$\bar{G}_m(\bar{s}) = \frac{1}{\bar{s}+1}, \; \bar{G}_p(\bar{s}) = \frac{1}{(1 \pm 10\%)\bar{s}+1} \tag{8.11}$$

The uncertainty is described by $l(\bar{w})$.

$$l(\bar{w}) = \max_{\Pi \in \pm 10\%} \left| \frac{(\Pi\bar{s}+1)}{\bar{s}+1} - 1 \right| \tag{8.12}$$

Figure 8.7(a) shows the magnitude of 20 combinations of the uncertain $l(\bar{w})$, together with the upper bound error weight given by the first order transfer function

$$W(s) = \frac{0.1s}{s+1} \tag{8.13}$$

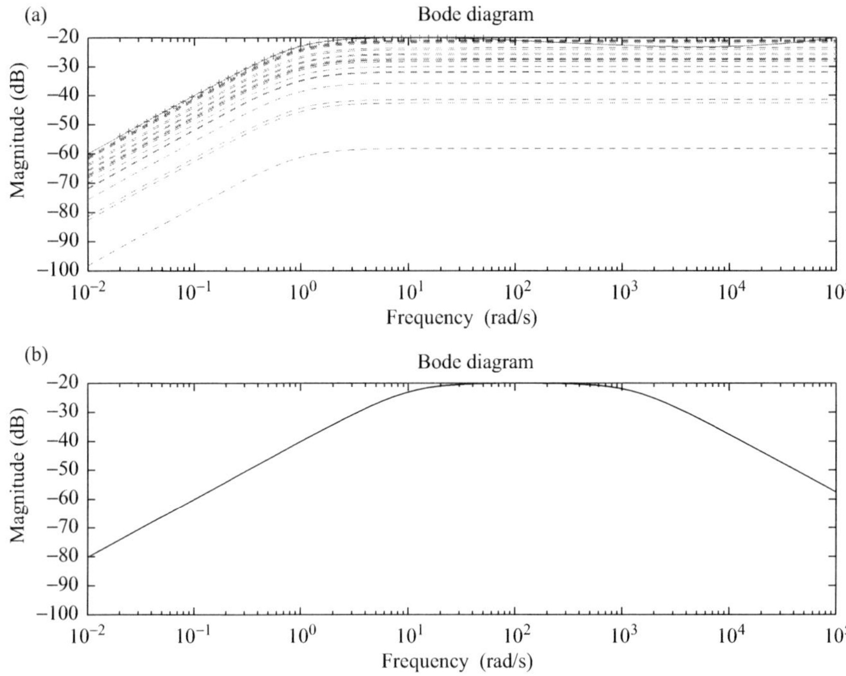

Figure 8.7 Relative error for 20 combinations of uncertain zero. '+' line is the magnitude of the first order weight

The condition of robust stability $\|W(s)T(s)\|_\infty < 1$, with $T(s) = G_m(s)C_m(s)(1 + G_m(s)C_m(s))^{-1}$, is accomplished as can be seen in Figure 8.7(b). As a result the controller $C_p(s)$ does guarantee robust stability when applied on the prototype under partial similarity.

References

[1] P. Balaguer, C. Pedret A. Ibeas, and S. Alcántara. 'Controller parameter dependence on model information through dimensional analysis.' In *48th Conference on Decision and Control, Shanghai*, 2009

[2] G.I. Barenblatt. *Dimensional Analysis*. Gordon and Breach Science Publishers, 1987

[3] S.P. Bhattacharyya, H. Chapellat, and L.H. Keel. *Robust Control. The Parametric Approach*. Prentice Hall, 1995

[4] S. Brennan and A. Alleyne. 'The Illinois roadway simulator: a mechatronic testbed for vehicle dynamics and control.' *IEEE/ASME Transactions on Mechatronics*, 5:349–359, 2000

[5] S. Brennan and A. Alleyne. 'Dimensionless robust control with applications to vehicles.' *IEEE Transactions on Control Systems Technology*, 13:624–630, 2005

[6] ecp Educational Control Products. *Industrial Plant Emulator Model 220*, http://www.ecpsystems.com/controls_emulator.htm, 2008

[7] M. Ghanekar, D.W.L. Wang, and G.R. Heppler. 'Scaling laws for linear controllers of flexible link manipulators characterized by nondimensional groups.' *IEEE Transactions on Robotics and Automation*, 13:117–127, 1997

[8] M. Higashimori, M. Harada, M. Yuya, I. Ishii, and M. Kaneko. 'Dimensional analysis based design on tracing type legged robots.' In *Proceedings of the 2005 IEEE International Conference on Robotics and Automation*, Barcelona, Spain, 2005

[9] M. Kaneko, S. Tachi, K. Tanie, and M. Abe. 'Basic study on similarity in walking machine from a point of energetic efficiency.' *IEEE Journal of Robostics and Automation*, RA-3:19–30, 1987

[10] S. Skogestad and I. Postlethwaite. *Multivariable Feedback Control. Analysis and Design*. Wiley, 1996

[11] T. Szirtes and P. Rózsa. *Applied Dimensional Analysis and Modeling*. Butterworth-Heinemann, 2007

[12] F.M. White. *Fluid Mechanics*. Mcgraw-Hill College, 4 Har/Dsk edition, 1998

[13] M. Zlokarnik. *Scale-up in Chemical Engineering*. Wiley-VCH, 2002

Chapter 9
Adaptive systems

9.1 Introduction

Adaptive control systems are used to overcome the problem of plant uncertainty due to, for instance, slowly varying time parameters [4]. One approach to adaptive control is gain scheduling, where an appropriate controller is chosen on the basis of scheduling variables that are continuously observed [2]. The complexity of a gain scheduling controller increases with the number of scheduling parameters. Dimensional analysis may reduce the problem complexity formulation by using a dimensionless representation that reduces the number of scheduling parameters, as shown in Reference 3.

Another common technique for adaptive systems is the model reference adaptive control (MRAC). In this case the controller parameters are continuously updated in order to keep a fixed closed loop performance, specified by a reference model [2]. Despite the fact that a constant reference model is common practice, better MRAC schemes can be achieved by considering time varying reference models during operation. The motivation for a time varying reference model can be better understood by means of an example. Consider the adaptive attitude control of a spacecraft. Due to fuel consumption the axis inertia moments diminish with time. On the other hand, when the fuel is refilled in the spacecraft (assume it can be done during normal operation) the spacecraft inertia increases. In this case it could not be sensible to have a constant reference model because:

- Due to inertia increase, it can be the case that too much fuel is expended or even actuators can be saturated if closed loop dynamics are too demanding.
- In case of inertia decrease it could be possible to achieve a faster closed loop performance level, more suited for landing maneuvers.

As a result it can be difficult to choose a valid reference model with suited dynamics for the whole spacecraft inertia operation range and it could lead either to conservative results when low inertia is present or to actuator saturation and fuel wasting under the high inertia condition.

A more convenient approach is to have a variable reference model depending on the spacecraft inertia. In this case it is possible to set faster closed loop dynamics when the inertia is small with no controller saturation. On the other hand, when the spacecraft has high inertia, the reference model dynamics are set to slower values in order to prevent actuator saturation.

In this chapter we show how dimensional analysis concepts are used to derive a new model reference adaptive control with the following features: (i) the reference model is no longer constant but time varying: (ii) there is a bounded control effort, measured as the shift between the current system poles and the reference model poles. Finally (iii) the reference model variation is no longer arbitrary but follows a similar pattern. In this way although the reference model is no longer constant, several properties of the reference model are still kept constant. In fact the reference model describes similar dynamics.

The similar model reference adaptive control (SMRAC) is one approach to MRAC schemes in which actuator wearing and saturation are concerns. The SMARC guarantees a bounded control effort during operation at the cost of modifying the reference model but maintaining output dimensionless properties constants.

9.2 Actuator limitations and dimensionally similar model reference

9.2.1 Control effort

Normally, the actuator limitations are set as bounds on the control signal magnitude or its rate of change. However, we introduce a new way of setting the actuator capacity by means of the control effort concept. The control effort c, as defined in Reference 1, is the shift of the poles from the open to the closed loop system

$$c = -\sum_{i=1}^{n} (b_i - a_i) \tag{9.1}$$

where n is the system order, and b_i and a_i are the closed loop poles and the open loop poles, respectively. Equation (9.1) can be rewritten as

$$c = B_n - A_n \tag{9.2}$$

with $B_n = -\sum_i^n b_i$ and $A_n = -\sum_i^n a_i$.

The relation between the closed loop poles and control effort can be classified accordingly to its sign as

- $c > 0$. In this case the closed loop poles are faster than the open loop poles.
- $c = 0$. The closed loop poles are equal to the open loop poles.
- $c < 0$. The closed loop poles are slower than the open loop poles.

In the MRAC the number of open loop poles (i.e. plant poles) is not necessarily equal to the number of closed loop poles (i.e. the model reference poles) because the denominators of the plant and the reference model are not necessarily equal. In this case a new definition of the control effort is required and stated as follows:

$$c = B_{n_m} - A_{n_p} \tag{9.3}$$

with $B_{n_m} = -\sum_i^{n_m} b_i$ and $A_{n_p} = -\sum_i^{n_p} a_i$, where b_i and a_i are poles of the closed loop and the poles of the open loop, respectively. In this case the above classification is no longer valid as the control effort can be equal to zero with closed loop poles distinct to the open loop poles. Anyway, it is true that any relative increase in the closed loop poles stability comes by an increase in the control effort, for constant A_{n_p}.

9.2.2 Similar model reference adaptive control

Dimensionally similar models

Recall that if two transfer functions, the model $G_m(s_m)$ and the prototype $G_p(s_p)$, are dimensionally similar, then they are related as

$$G_p(s_p) = S_k S_t^l G_m(S_t s_p) \tag{9.4}$$

where S_k and S_t are the gain and time scaling factors, respectively, and l is the number of plant integrators. In the same way, given a nominal transfer function $G_0(s_0)$, any transfer function is derived from $G_0(s_0)$ by

1. Multiplying $G_0(s_0)$ by the scaling factors $S_k S_t^l$
2. Substituting the variable s_0 by $S_t s$, that is $s_0 \to S_t s$

is a transfer function dimensionally similar to $G_0(s_0)$, for any $S_k > 0$ and $S_t > 0$. In general, given an nth-order nominal transfer function $G_0(s_0)$ with relative degree equal to n^*

$$G_0(s_0) = \frac{k_0(s_0 + a_{0n+1}) \dots (s_0 + a_{0n+m})}{(s_0 + a_{01}) \dots (s_0 + a_{0n})} \tag{9.5}$$

$G(s)$ is similar to $G_0(s)$ if

$$G(s) = \frac{k(s + a_{n+1}) \dots (s + a_{n+m})}{(s + a_1) \dots (s + a_n)} \tag{9.6}$$

with

$$\begin{aligned} k &= S_k S_t^{l-n^*} k_0 \\ a_i &= S_t^{-1} a_{0i}, \quad i \in [1, \dots, n+m] \\ S_k &> 0 \\ S_t &> 0 \end{aligned} \tag{9.7}$$

In words, the $G(s)$ gain is scaled by the gain scale factor S_k and by n^* times the inverse time scaling factor. The poles and zeroes of $G(s)$ are scaled by the inverse of the time scaling factor. Note that $S_t > 1$ means that the system $G(s)$ is slower than $G_0(s)$. On the contrary, $S_t < 1$ means that the system $G(s)$ is faster than $G_0(s)$.

Now two important lemmas are established for similar transfer functions.

Lemma 9.1. *If transfer function $G_0(s)$ is coprime, then any transfer function $G(s)$ obtained by similar transformation from $G_0(s)$ is also coprime.*

Proof. The result follows because similar transformation scales all poles and zeroes by the same positive scaling factor S_t^{-1}. □

Lemma 9.2. *If $G_0(s_0)$ is a strictly positive real (SPR) transfer function, then any transfer function $G(s)$ obtained by similar transformation from $G_0(s_0)$ is also SPR.*

Proof. By definition (see for instance Reference 5), $G_0(s_0)$ is SPR if and only if (i) $G_0(s_0)$ is a strict stable transfer function and (ii) the real part of $G_0(s_0)$ is strict positive along the jw_0 axis, that is, $\forall w_0 \geq 0\ Re[G_0(jw_0)] > 0$. If $G(s)$ is derived from $G_0(s_0)$ by a similarity transformation it follows from the preceding discussion that $G(s) = S_k S_t^{l-n^*} G_0(s_0 \to S_t s)$.

The multiplication by the constant $S_k S_t^{l-n^*}$ keeps the poles unaltered. The variable substitution $s_0 \to S_t s$ means that the poles (and also the zeroes) of $G(s)$ are scaled by the inverse of the time scaling factor. As $S_t > 0$ the stability properties are unchanged. As a result if $G_0(s_0)$ is strictly stable, so is $G(s)$.

By assumption $\forall w_0 \geq 0\ Re[G_0(jw_0)] > 0$. It then follows that $\forall S_t w_0 \geq 0$ $Re[G_0(jS_t w_0)] > 0$, for $S_t > 0$. Moreover, as $S_k > 0\ \forall S_t w_0 \geq 0\ S_k S_t^{l-n^*} Re[G_0(jS_t w_0)] > 0$, which by similarity is analogous to $\forall w \geq 0\ Re[G(jw)] > 0$. □

Adaptive scheme

Let an nth-order linear time invariant (LTI) plant with input–output pair $\{u_p(t), y_p(t)\}$ be described by the transfer function

$$W_p(s) = k_p \frac{Z_p(s)}{R_p(s)} \tag{9.8}$$

$Z_p(s)$ is an m_p degree monic polynomial and $R_p(s)$ is an n_p degree monic polynomial, with $n_p \geq m_p$. It is assumed that $Z_p(s)$ and $R_p(s)$ are coprime.

Let a linear time varying (LTV) model reference transfer function with input–output pair $\{r(t), y_m(t)\}$ be defined by

$$W_m(s,t) = k_m(t) \frac{Z_m(s,t)}{R_m(s,t)} \tag{9.9}$$

with

$$\begin{aligned} Z_m(s,t) &= (s + a_{m(n_m+1)}(t))(s + a_{m(n_m+2)}(t)) \dots (s + a_{m(n_m+m_m)}(t)) \\ R_m(s,t) &= (s + a_{m1}(t))(s + a_{m1}(t)) \dots (s + a_{mn_m}(t)) \end{aligned} \tag{9.10}$$

$Z_m(s,t)$ is an m_m degree monic polynomial and $R_m(s,t)$ is an n_m degree monic polynomial, with $n_m \geq m_m$. Let $W_m(s, t_0)$ be the initial reference model (i.e. $t \geq t_0$)

with $Z_m(s, t_0)$ and $R_m(s, t_0)$ coprime. $W_m(s, t)$ varies in a similar manner with respect to $W_m(s, t_0)$, to be precisely defined in the following section.

Our objective is to design an adaptive control system with an adaptive controller with *bounded control effort* by varying the model reference in a *dimensionally similar* manner, in order to keep certain input–output properties constant. It is required that the adaptive control guarantees the boundedness of the tracking error and its convergence to zero, where $r(t)$ is assumed to be bounded, piecewise continuous and persistently exciting of appropriate order to guarantee the converge of the plant-identified parameters to their real values.

Figure 9.1 shows the proposed adaptive scheme. It is an indirect MRAC with time varying model reference W_m adapted by the control effort of the closed loop in such a way that the control effort is bounded by c^{max}.

The assumptions regarding the plant and the reference model are summarized as follows:

- *Plant assumptions*
  - (P1) Z_p is a Hurwitz polynomial.
  - (P2) The relative degree $n_p^* = n_p - m_p$ is known.
  - (P3) A minimum value of the high frequency gain k_p^{min} is known.
  - (P4) An upper bound n of n_p is known.

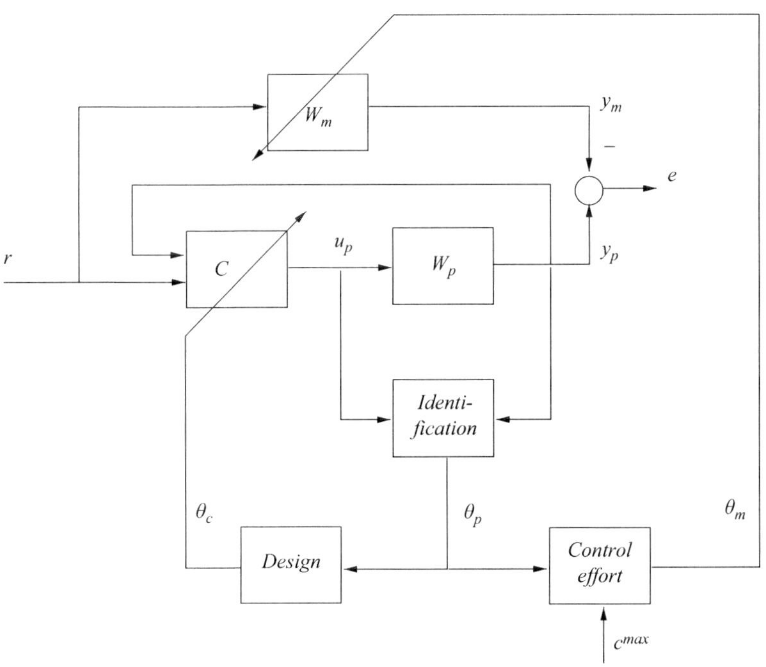

Figure 9.1 Similar model reference adaptive control (SMRAC)

- *Reference model assumptions*
  (M1) The relative degree $n_m^* = n_m - m_m$ is greater than or equal to n_p^*, that is, $n_m^* \geq n_p^*$.
  (M2) At any time instant, $W_m(s,t)$ is dimensionally similar to $W_m(s,t_0)$ for $t \geq t_0$ (i.e. the variations of $W_m(s,t)$ are not arbitrary but follow a similar pattern).

Regarding plant assumptions, P2 and P4 are required to establish the existence of a controller that solves the closed loop matching problem. Assumptions P1 and P3 guarantee that the generated adaptive law is stable. On the one hand P1 prevents unstable pole-zero cancelations. On the other hand P3 guarantees a bounded control action. In fact, perfect control requires the feedforward term k_m/k_p, which can become unbounded if k_p is not limited below. Finally reference model assumption M1 is necessary to achieve perfect tracking, while assumption M2 is a design property enabling the SMRAC approach.

9.3 SMRAC for first order plants

In order to make the main ideas of the SMRAC approach clear, we first develop in detail the SMRAC for first order linear time-invariant plant. In Section 9.4 we generalize the results for arbitrary order plants.

Consider the first order LTI plant

$$\dot{y}_p(t) = a_p y_p(t) + k_p u(t) \tag{9.11}$$

where $u(t)$ and $y_p(t)$ are the input and output of the plant, respectively. The plant parameters a_p and k_p are constant or slowly varying, real, and unknown. The plant is controllable (i.e. $k_p \neq 0$) and it is not necessarily stable (i.e. values of $a_p > 0$ are allowed).

The reference model is a similar time varying system described by

$$\dot{y}_m(t) = a_m(t) y_m(t) + k_m(t) r(t) \tag{9.12}$$

where $r(t)$ and $y_m(t)$ are the reference input and output, respectively, of the reference model. The parameters $a_m(t)$ and $k_m(t)$ are known time varying scalar constants with $a_m(t) < 0$ (i.e. the reference model is stable). The reference model parameters variation is not arbitrary but similar, in such a way that, at any time instant the relation between the model reference parameters is given by

$$\begin{aligned} a_m(t) &= S_t(t)^{-1} a_{0m} \\ k_m(t) &= S_k(t) S_t(t)^{-1} k_{0m} \end{aligned}$$

with $S_k > 0$, $S_t > 0$ the gain and time scaling factors, respectively, and a_{0m} and k_{0m} initial or nominal values of the reference model.

The time scaling factor S_t is modified regarding the currently estimated control effort $c(t)$ (calculated using the parameter estimation $\hat{a}_p(t)$)

$$c(t) = -(a_{0m} - \hat{a}_p(t)) \tag{9.13}$$

as follows

$$S_t(t) = \begin{cases} 1 & \text{if } c(t) \leq c^{max} \\ \dfrac{a_{0m}}{\hat{a}_p(t) - c^{max}} & \text{if } c(t) > c^{max} \end{cases} \tag{9.14}$$

If $c(t) \leq c^{max}$ then the model reference W_m is kept constant. On the contrary, if $c(t) > c^{max}$, the time scale factor is modified. As a result, it is guaranteed that the control effort is bounded by c^{max}, independently of the value of $\hat{a}_p(t)$, varying due to either estimation or slow plant parameters changes.

It is assumed that if $a_p > 0$ (i.e. the plant is unstable) then $c^{max} > a_p(t)$, because for stabilization purposes it is required that $a_m = c - a_p < 0$. If the system is stable $a_p < 0$ then it is just required that $c^{max} \geq 0$. In any case $S_t > 0$.

The gain scaling factor S_k can be kept constant and equal to unity because its value does not affect the control effort. It can also be modified if a variable model reference steady state gain is desired.

Our objective is to make the control error $e_c = y_p - y_m$ tend to zero with time with bounded control effort, using an indirect control scheme in which the model reference, when required, is no longer constant but varies in a similar way. The control input is generated as

$$u(t) = \theta(t)y_p(t) + k(t)r(t) \tag{9.15}$$

where $\theta(t)$ and $k(t)$ are the controller parameters to be adjusted with an appropriate adaptive law that accomplishes the above mentioned objectives. Following classical indirect MRAC, the controller parameters that provide the desired model reference behavior of the closed loop are given by

$$\begin{aligned} \theta(t) &= \frac{a_m - a_p(t)}{k_p} \\ k(t) &= \frac{k_m(t)}{k_p} \end{aligned} \tag{9.16}$$

Because the plant parameters are unknown, the controller parameters are calculated using the plant parameters estimates

$$\begin{aligned} \theta(t) &= \frac{a_m(t) - \hat{a}_p(t)}{\hat{k}_p(t)} \\ k(t) &= \frac{k_m(t)}{\hat{k}_p(t)} \end{aligned} \tag{9.17}$$

The parameter adaptation law

$$\dot{\hat{a}}_p = \gamma_1 y_p(t) e(t) \tag{9.18}$$

$$\dot{\hat{k}}_p = \gamma_2 u_p(t) e(t) \tag{9.19}$$

with γ_1 and γ_2 positive constants, together with

$$a_m(t) = S_t^{-1} a_{0m}$$
$$k_m(t) = S_k S_t^{-1} k_{0m}$$

and

$$\dot{S}_t(t) = \begin{cases} 0 & \text{if } c(t) \leq c^{max} \\ \dfrac{-a_{0m}}{(\hat{a}_p(t) - c^{max})^2}\dot{\hat{a}}_p & \text{if } c(t) > c^{max} \end{cases}$$

yields a stable adaptive parameter estimation with bounded control effort. Note that the controller parameter $\theta(t)$ is bounded by c^{max}/k_p^{min}.

The stability of the SMRAC is established as follows. The identification model is described by Reference 5

$$\dot{\hat{y}}_p(t) = a_m(t)\hat{y}_p + (\hat{a}_p(t) - a_m(t))y_p(t) + \hat{k}_p(t)u(t) \tag{9.20}$$

with $a_m(t) < 0$ for stability of the reference model. The above model yields the following identification error (output error): $e_i = y_p(t) - \hat{y}_p(t)$. The error variation with time is

$$\dot{e}_i = \dot{\hat{y}}_p(t) - \dot{y}_p(t) = a_m(t)e_i(t) + (-\hat{a}_p + a_p)y_p(t) - (-\hat{k}_p - k_p)u(t) \tag{9.21}$$

which can be written as

$$\dot{e}_i = a_m(t)e_i(t) - \tilde{a}_p y_p(t) - \tilde{k}_p u(t) \tag{9.22}$$

where $\tilde{a}_p = \hat{a}_p - a_p$ and $\tilde{k}_p = \hat{k}_p - k_p$ are the plant parameters error.

Consider the following Lyapunov function:

$$V = \frac{1}{2}(e^2 + \gamma_1 \tilde{a}_p^2 + \gamma_2 \tilde{k}_p^2) \tag{9.23}$$

with time derivative given by

$$\dot{V} = a_m(t)e^2 + (\dot{\tilde{a}}_p - y_p e)\tilde{a}_p + (\dot{\tilde{k}}_p - ue)\tilde{k}_p \tag{9.24}$$

If the plant parameters are adjusted using the adaptive laws ((9.18) and (9.19)) the stability of the adaptive scheme follows because

$$\dot{V} = a_m(t)e^2 < 0 \tag{9.25}$$

To show this consider the parameter $a_m(t)$:

$$a_m(t) = \begin{cases} a_{0m} & \text{if } c \leq c^{max} \\ S_t^{-1} a_{0m} & \text{if } c > c^{max} \text{ with } S_t > 0 \end{cases} \tag{9.26}$$

In both cases $a_m(t) < 0$ yielding system stability because the time scaling factor S_t as calculated in (9.14) is always positive.

9.4 SMRAC for arbitrary order plants

9.4.1 SMRAC control scheme

Consider the following n_p order LTI plant:

$$W_p(s) = k_p \frac{Z_p(s)}{R_p(s)} \tag{9.27}$$

and the n_m order LTV reference model transfer function

$$W_m(s,t) = k_m(t) \frac{Z_m(s,t)}{R_m(s,t)} \tag{9.28}$$

together with the assumptions established in Section 9.2.2.

The reference model changes with time following a similar pattern. As a result the parameters of the time varying polynomials $Z_m(s,t)$ and $R_m(s,t)$, defined as

$$\begin{aligned} Z_m(s,t) &= (s + a_{m(n_m+1)}(t))(s + a_{m(n_m+2)}(t)) \ldots (s + a_{m(n_m+m_m)}(t)) \\ R_m(s,t) &= (s + a_{m1}(t))(s + a_{m1}(t)) \ldots (s + a_{mn_m}(t)) \end{aligned} \tag{9.29}$$

are given by

$$\begin{aligned} a_{mi}(t) &= S_t^{-1} a_{0mi} \quad i \in [1, \ldots, n_m + m_m] \\ k_m(t) &= S_k S_t^{-1} k_{0m} \end{aligned} \tag{9.30}$$

with S_k, S_t the gain and time scaling factors, respectively, and a_{0mi} and k_{0m} initial values of the reference model (i.e. nominal reference model). We keep, without loss of generality, the gain scaling factor constant and equal to 1, that is, $S_k = 1$, in order to keep a constant reference model high frequency gain.

The estimated parameters of the plant $\hat{p} = [\hat{c}_0, \hat{\bar{c}}^T, \hat{d}_0, \hat{\bar{d}}^T]^T$, as parameterized by the observer shown in Figure 9.2, are continuously updated by the following adaptive law:

$$\dot{\hat{k}} = -sgn(k_p)\epsilon_1 \bar{\zeta}_{\bar{u}} \tag{9.31}$$

$$\dot{\hat{p}} = -\epsilon_1 \bar{\zeta}_o \tag{9.32}$$

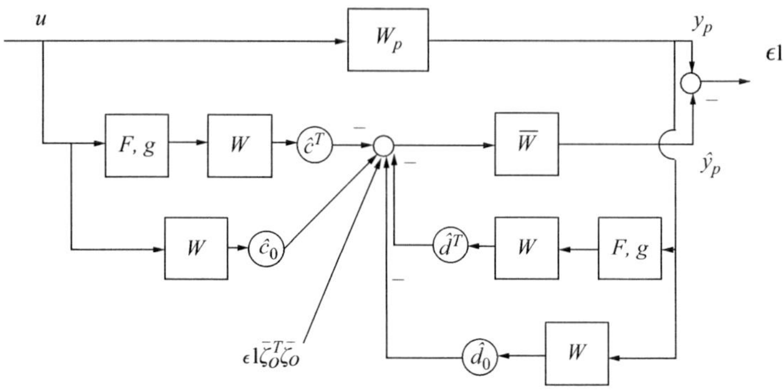

Figure 9.2 Observer parameterization on the SMRAC

where $\hat{k} = \hat{c}_0^{-1}$, $\hat{\bar{p}} = [\hat{\bar{c}}^T, \hat{d}_0, \hat{\bar{d}}^T]^T$, $\epsilon_1 = y_p - \hat{y}_p$, and ζ_o, $\zeta_{\bar{u}}$ measurable signals, which for the sake of clarity are defined in Section 9.5.

The current control effort estimation is given by

$$c(t) = A_{m0}^{n_m} - \hat{A}_p^{n_p}(t) \tag{9.33}$$

with $A_{m0}^{n_m} = -\sum_{i=1}^{n_m} a_{0i}$ (i.e. a_{0i} poles of $W_{0m}(s)$) and $\hat{A}_p^{n_p} = -\sum_{i=1}^{n_p} \hat{a}_{pi}$ (i.e. $\hat{a}_{pi}$ poles of $\hat{W}_p(s)$). The poles $\hat{A}_p^{n_p}$ are calculated by means of the estimated parameters $\hat{p}$ and the representation shown in Figure 9.2.

The time scaling factor S_t is modified regarding the currently estimated control effort $c(t)$ as follows:

$$S_t(t) = \begin{cases} 1 & \text{if } c(t) \leq c^{max} \\ \dfrac{A_{m0}^{n_m}}{c^{max} + \hat{A}_p^{n_p}(t)} & \text{if } c(t) > c^{max} \end{cases}$$

If $c(t) \leq c^{max}$ then the reference model W_m is kept constant and equal to the nominal model $W_m(s, t_0)$. On the contrary, if $c(t) > c^{max}$, the time scale factor is modified. As a result, it is guaranteed that the control effort is bounded by c^{max}, independently of the value variations of $\hat{A}_p^{n_p}$, due to either estimation or slow plant parameters changes.

Assumption 9.1. *In case of W_p stable (i.e. all $a_{pi} < 0$ $i \in [1, \ldots, n_p]$), then $c^{max} \geq 0$. This assures that the closed loop poles $A_m^{n_m}$ are, at least, equal to the open loop ones.*

Assumption 9.2. *In case of W_p unstable (i.e. there is at least one $a_{pi} > 0$ $i \in [1, \ldots, n_p]$), then $c^{max} > -A_{pu}^{n_{pu}}$ with $A_{pu}^{n_{pu}} = -\sum_1^{n_{pu}} a_{pu}$ and $a_{pu} > 0$, that is the unstable poles of W_p. This assumption is necessary in order to be able to shift the open loop unstable poles to closed loop stable poles. Analytically it can be shown by rewriting (9.33) as $c = A_m^{n_m} - A_{ps}^{n_{ps}} - A_{pu}^{n_{pu}}$ separating the stable and the unstable poles. It* *then follows that, disregarding the plant stable poles, $A_m^{n_m} = c + A_{pu}^{n_{pu}}$. If $c^{max} > A_{pu}^{n_{pu}}$ then $A_m^{n_m} > 0$, thus W_p is stable.*

The consequence of both assumptions is that the scaling factor S_t is always positive. Finally the control action that maintains a bounded control effort is calculated as

$$u = \hat{k}(r + \hat{\bar{p}}^T) \tag{9.34}$$

9.4.2 SMRAC stability analysis

Consider the observer

$$\begin{aligned} \dot{w}_1 &= Fw_1 + gu_p, \ w_1(0) = 0 \\ \dot{w}_2 &= Fw_2 + gy_p, \ w_2(0) = 0 \end{aligned} \tag{9.35}$$

with $\bar{w}_o = [w_{1o}^T, y_p, w_{2o}^T]^T$ and $\hat{\bar{p}} = [\hat{\bar{c}}^T, \hat{d}_0, \hat{\bar{d}}^T]^T$.

We modify a standard observer yielding the augmented observer presented in Figure 9.2, with output

$$\hat{y}_p = \overline{W}(\hat{c}_0 Wu - \hat{\bar{p}}^T WIw_0 + \epsilon_1 \bar{\zeta}_o^T \bar{\zeta}_o) \tag{9.36}$$

and $W_m = \overline{W}W$[1] with $\overline{W}$ SPR and $\bar{\zeta}_o = W\bar{w}_o$ the filtered observer state. The control action is given by $u = \hat{k}(r + \hat{\bar{p}}^T)$, resulting

$$\hat{y}_p = \overline{W}(\hat{c}_0 W\hat{k}(r + \hat{\bar{p}}^T) - \hat{\bar{p}}^T WIw_0 + \epsilon_1 \bar{\zeta}_o^T \bar{\zeta}_o) \tag{9.37}$$

which can be further simplified by linearity, yielding

$$\hat{y}_p = y_m + \overline{W}\epsilon_1 \bar{\zeta}_o^T \bar{\zeta}_o \tag{9.38}$$

The augmented observer error $\epsilon_1 = y_p - \hat{y}_p$ is then written as

$$\epsilon_1 = y_p - \hat{y}_p = y_p - y_m - \overline{W}\epsilon_1 \bar{\zeta}_o^T \bar{\zeta}_o \tag{9.39}$$

Now consider the tracking error $e_1 = y_p - y_m$ (see Figure 9.3). The control action $u = \hat{k}(r + \hat{\bar{p}}^T)$ can be rewritten as $u = (k + \psi_1)(r + \bar{p}^T + \bar{\Delta}^T \bar{w}_o)$ where $\psi_1 = \hat{k} - k$ and $\bar{\Delta} = \hat{\bar{p}} - \bar{p} = [\hat{c}^T - c^T, \hat{d}_o - d_o, \hat{d}^T - d^T]$ are the parameter errors. The tracking error is

$$e_1 = \frac{k_p}{k_m} W_m(k\bar{\Delta}^T \bar{w}_o + \psi_1(r + \bar{p}\bar{w}_o + \bar{\Delta}\bar{w}_o)) \tag{9.40}$$

[1] The dependence of W_m, $\overline{W}$ and W on time is dropped for readability purposes. It is written explicitly when needed.

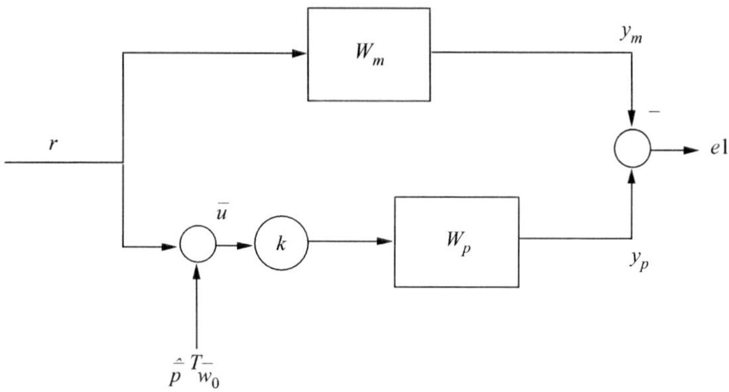

Figure 9.3 Reference model and plant error

from Figure 9.3, $\bar{u} = r + \bar{p}\bar{w}_o + \bar{\Delta}\bar{w}_o$ and for perfect tracking $k = k_m/k_p$, resulting

$$\begin{aligned} e_1 &= \frac{k_p}{k_m} W_m(k\bar{\Delta}^T \bar{w}_o + \psi_1 \bar{u}) \\ &= \overline{W}\left(\frac{k_p}{k_m} W \psi_1 \bar{u} + W \bar{\Delta}^T \bar{w}_o\right) \\ &= \overline{W}\left(\frac{k_p}{k_m} \psi_1 \bar{\zeta}_{\bar{u}} + \bar{\Delta}^T \bar{\zeta}_o\right) \end{aligned} \tag{9.41}$$

Finally, recalling that $\zeta_o = W\bar{w}_o$ is the filtered observer state and $\zeta_{\bar{u}} = W\bar{u}$ is the filtered input $\bar{u}$, ϵ_1 can be written as

$$\epsilon_1 = \overline{W}\left(\frac{k_p}{k_m} \psi_1 \bar{\zeta}_{\bar{u}} + \bar{\Delta}^T \bar{\zeta}_o - \epsilon_1 \bar{\zeta}_o^T \bar{\zeta}_o\right) \tag{9.42}$$

The following adaptation laws are proposed due to the SPR of $\overline{W}$:

$$\dot{\psi}_1 = \dot{\hat{k}} = -sgn(k_p)\epsilon_1 \bar{\zeta}_{\bar{u}} \tag{9.43}$$

$$\dot{\bar{\Delta}} = \dot{\hat{\bar{p}}} = -\epsilon_1 \bar{\zeta}_o \tag{9.44}$$

Now it is shown that the adaptive system is stable. Recall that (9.42) can be rewritten in state space form as

$$\dot{\epsilon} = A_1 \epsilon + b_1 \left(\frac{k_p}{k_m} \psi_1 \bar{\zeta}_{\bar{u}} + \bar{\Delta}^T \bar{\zeta}_o - \epsilon_1 \bar{\zeta}_o^T \bar{\zeta}_o\right) \tag{9.45}$$

$$\epsilon_1 = h_1^T \epsilon \tag{9.46}$$

with $h_1^T(sI - A_1)^{-1} b_1 = \overline{W}$ SPR.

We assume a Lyapunov candidate equation of the form

$$V = \epsilon^T P\epsilon + \frac{|k_p|}{k_m}\psi_1^2 + \bar{\Delta}^T\bar{\Delta} \tag{9.47}$$

which has a time derivative equal to

$$\begin{aligned}\dot{V} &= \epsilon^T(A_1^T P + PA_1)\epsilon + 2\epsilon^T Pb_1\left(\frac{k_p}{k_m}\psi_1\bar{\zeta}_{\bar{u}} + \bar{\Delta}^T\bar{\zeta}_o - \epsilon_1\bar{\zeta}_o^T\bar{\zeta}_o\right)\\ &\quad - 2\frac{k_p}{k_m\psi_1\epsilon_1}\bar{\zeta}_{\bar{u}} - 2\epsilon_1\bar{\Delta}^T\bar{\zeta}_o\end{aligned} \tag{9.48}$$

As before, since $\overline{W}$ is SPR $\forall t$ due to the similarity transformation, it then follows by the Kalman–Yakubovich lemma that $A_1^T(t)P(t) + P(t)A_1(t) = -Q(t)$ with $Q(t)$ positive definite $\forall t$ and that $P(t)b_1(t) = h_1(t)$ also $\forall t$. Hence,

$$\dot{V} = -\epsilon^T Q(t)\epsilon - 2\epsilon_1^2\bar{\zeta}_o^T\bar{\zeta}_o \leq 0 \tag{9.49}$$

so that ϵ, ψ_1, and Δ are bounded and ϵ, $\dot{\psi}_1$, and $\dot{\Delta}$ are also square integrable functions. It then follows by signal growth rates arguments (see for instance Reference 5) that all signals are uniformly bounded, then $lim_{t\to\infty}\epsilon_1(t) = 0$ which implies that $lim_{t\to\infty}e_1(t) = 0$. However, the convergence of the parameter estimates to the real parameter values cannot be guaranteed because it depends on the persistently exciting nature of signal r.

9.4.3 SMRAC operation modes

So far we have derived the SMRAC with bounded control effort, meanwhile maintaining a nominal reference model W_{0m} as constant as possible. In fact, the variations of the nominal reference model are only performed when the control effort is about to exceed.

On the other hand we can easily devise a scheme with constant control effort during the whole operation range, equal to the maximum allowable control effort $c^{max}(t)$. In this case we are constantly modifying the reference model $W_m(t)$ and the time scaling factor $S_t(t)$ is modified as follows:

$$S_t(t) = \frac{A_{m0}^{n_m}}{c^{max} + \hat{A}_p^{n_p}} \tag{9.50}$$

with no control effort estimation required. Note that in this case an indirect approach is no longer mandatory.

Note that by definition, $S_t(t)$ is always positive and now, it is possible that $S_t(t) > 1$ because we are demanding a reference model 'faster' than W_{0m}. In any case, the stability is guaranteed and their selection follows the user desires. In the first case the reference model is kept as constant as possible and only modified when the estimated control effort is violated. In the second case the reference model is continuously modified in order to have a constant control effort operation.

9.5 Application example

We consider the system describing angular rigid body dynamics relating the angular velocity $\dot{\theta}$ to the applied torque T. The transfer function is

$$G(s) = \frac{\dot{\theta}(s)}{T(s)} = \frac{J^{-1}}{s + C/J} \tag{9.51}$$

where J is the inertia and C the viscous damping. We assume that the damping is constant and equal to unity (i.e. $C = 1$) but the inertia is time varying $J \in [0.5, 2]$ as follows:

$$J(t) = \begin{cases} 0.5 \quad \text{for } t \in [0\text{–}150] \\ 0.5 + \dfrac{1.5}{100}(t - 150) \quad \text{for } t \in [150\text{–}250] \\ 2 \quad \text{for } t \in [250\text{–}350] \end{cases}$$

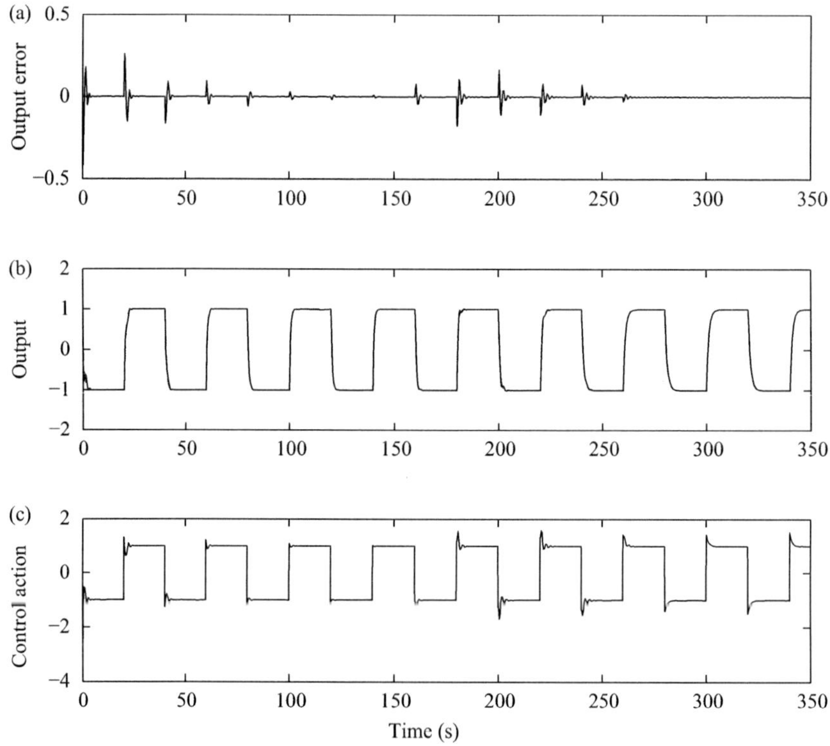

Figure 9.4 Signals

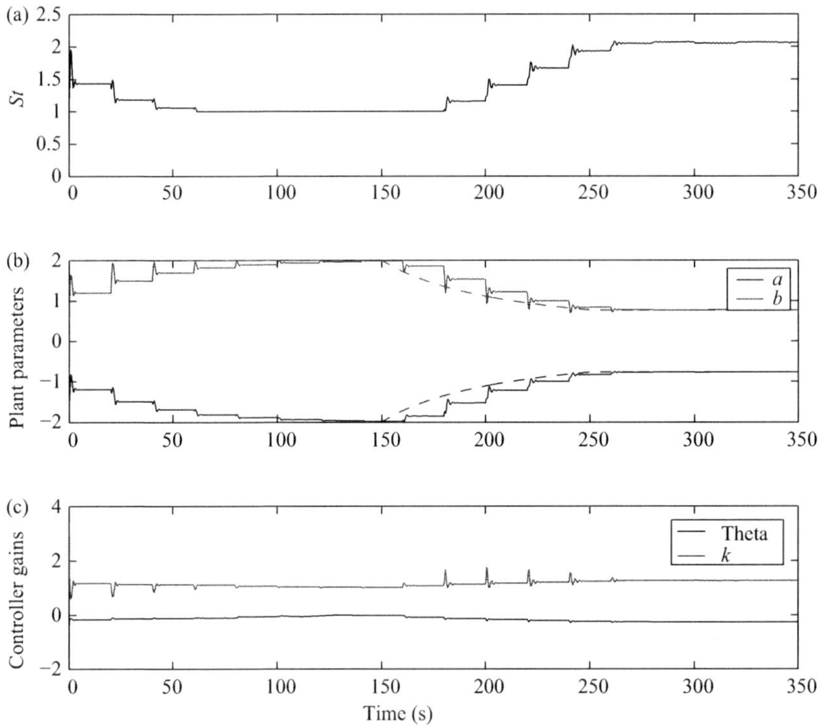

Figure 9.5 Parameters

The nominal values considered for the model reference model are $a_0 = -2$ and $k_0 = 2$, so

$$W_m(s, t_0) = \frac{2}{s+2} \tag{9.52}$$

and the reference model is time varying as a function of S_t as follows:

$$W_m(s, t) = \frac{S_k(2/S_t)}{s + 2/S_t} \tag{9.53}$$

In the example we keep $S_k = 1$, $\forall t$ in order to keep the steady-state gain unchanged. The maximum control effort bound selected is $c^{max} = 0.2$. Note in this case that, due to plant stability, no lower bound on c^{max} is required for stabilization purposes, so the only requirement is $c^{max} \geq 0$. The reference signal r is a square wave of zero mean value, amplitude equal to 1 and period 40 s.

Figure 9.4(a) presents the output error evolution. It tends to zero with time as expected. During the time interval $t \in [0\text{–}150]$ the output error is due to initial error parameters estimation, while during the time interval $t \in [150\text{–}250]$ the error is mainly due to slow parameter variation tracking. The identified parameters are shown in Figure 9.5(b) together with the real parameters evolution.

Next Figure 9.5(a) presents the model reference time scaling factor S_t evolution. The scaling factor S_t is equal to unity for $t \in [60–180]$. During that period the reference model is equal to the nominal model $W_m(s, t_0)$ previously described. However, during other operation periods the scaling factor reaches values greater than unity, so the reference model becomes slower, first due to estimation error and second due to inertia variation. This can be verified by Figure 9.4(b). It can also be seen in Figure 9.4(c) that the control action increases because inertia is very mild. Figure 9.5(c) shows also the time evolution of controller parameters $\theta(t)$ and $k(t)$. The $\theta(t)$ value is the feedback and the $k(t)$ is the feedforward.

Now we compare the evolution of two SMRAC schemes with distinct control effort bounds, $c^{max} = 0.2$ and $c^{max} = 2$. Figure 9.6(a) shows the model reference scaling factor S_t for both control effort bounds. It can be seen that although for $c^{max} = 0.2$ the S_t takes values greater than unity as seen before, for $c^{max} = 2$ the S_t value is equal to unity for all t. It thus follows that the reference model is kept constant and the results are equal to a standard MRAC.

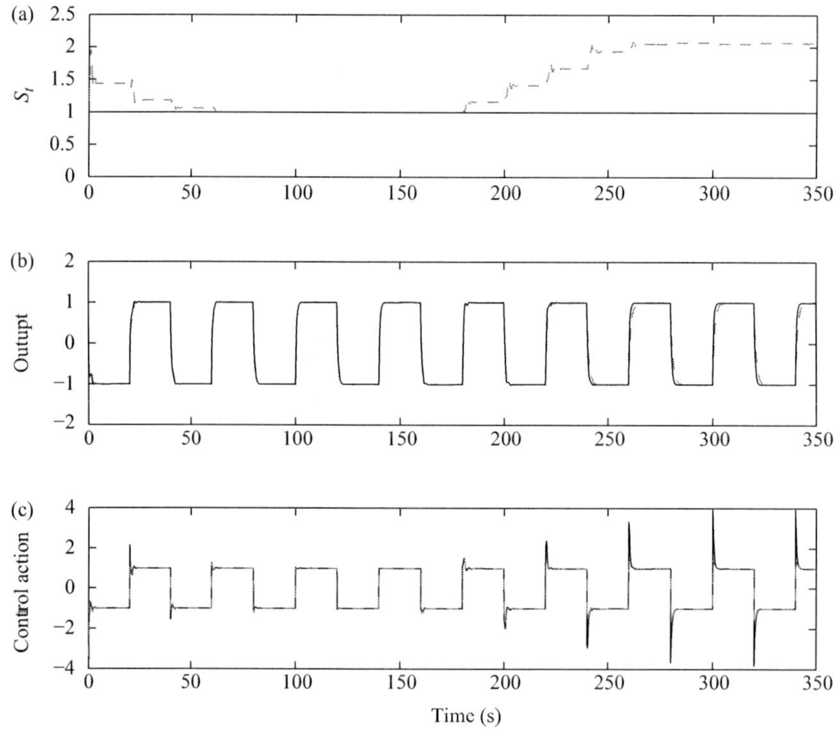

Figure 9.6 Comparison of responses with distinct control effort. Dashed line $c^{max} = 0.2$ and solid line $c^{max} = 2$

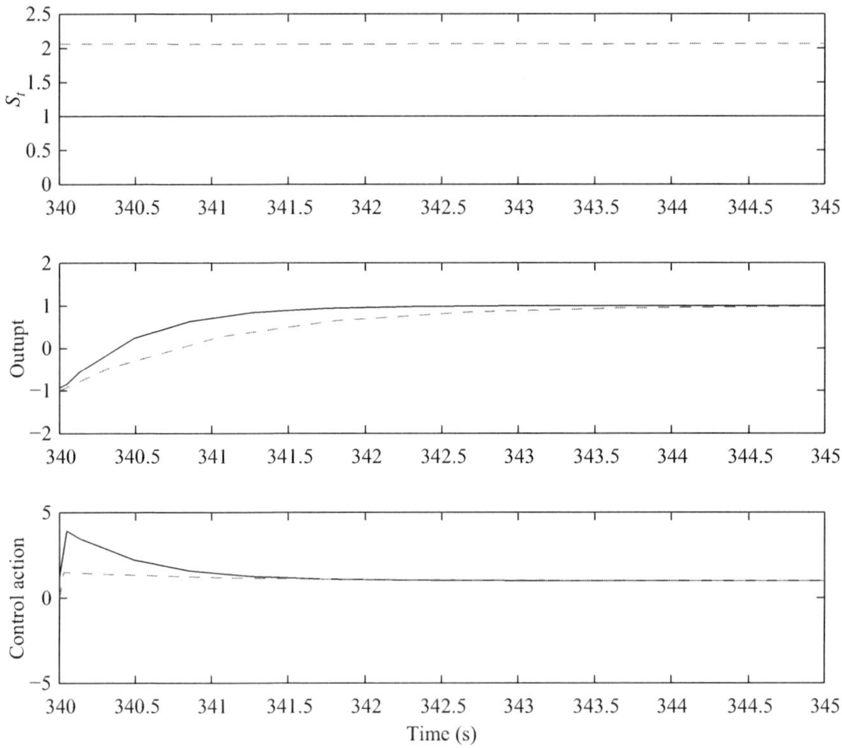

Figure 9.7 Comparison details

The comparison of outputs and control actions for both control effort bounds is presented in Figure 9.6(b) and (c). As expected, the SMRAC scheme with $c^{max} = 0.2$ shows a time response slower than the scheme with $c^{max} = 2$ when the inertia increases. However, the slight increase in the time response comes at the cost of higher control actions as seen in Figure 9.6(c). In fact the control action for $c^{max} = 2$ shows values around 4 units (spikes) while the control action for $c^{max} = 0.2$ is around 1 unit. Details of the presented plots are presented in Figure 9.7. The comparison shows how the control effort bound is useful to keep the control action bounded.

References

[1] P. Albertos and M. Olivares. 'Time delay limitations in control implementations.' In *European Control Conference*, Karlsrhue, 1999

[2] K.J. Astrom and B. Wittenmark. *Adaptive Control*. Addison-Wesley Publishing Company, 1989

[3] H. Hailu. *Dimensional tranformation: a novel method for gain-scheduling and robust control*. PhD thesis, The Pennsylvania State University, 2006
[4] I.D. Landau. 'From robust control to adaptive control.' *Control Engineering Practice*, 7:1113–1124, 1999
[5] K.S. Narendra and A.M. Annaswamy. *Stable Adaptive Systems*. Dover Publications, Inc., 1989

Index

actuator limitations 122–3
adaptive control systems 121–37
  see also similar model reference adaptive control (SMRAC)
arbitrary order plants, SMRAC for 129–33
  control scheme 129–31
  operation modes 133
  stability analysis 131–3
arithmetic of dimensions 12
  integral 12
  power 12
  product 12
  quotient 12

bandwidth: *see* sensitivity function bandwidth
Buckingham pi theorem 3, 14–16, 18
  see also dimensionless representation

closed loop transfer functions 83–6
  comparative framework, PID tuning rules 102
  dimensionless 85–6
  loop transfer function GC 83–5
comparative framework, PID tuning rules
  application example 102–3
  characterization 103
  closed loop transfer functions 102
  control action variability 106
  elements 98–9
  framework 99–100
  indexes 104
  integral absolute error 104, 105
  integral errors 100–2
  loop transfer function GC 102
  overview 97
  selecting tuning rules 106–7
  sensitivity function bandwidth 103, 104
  sensitivity peak function 104, 105
continuous time control 111
control effort 122–3
control engineering 8–9
  see also comparative framework, PID tuning rules
control of dimensionally similar systems 109–11
  complete similarity 111–13
  experimental case study 115–19
  overview 109–10
  partial similarity 113–15
control theory 8
  see also homogeneity, of PID tuning rules

decentralized controller: *see* reduced effective transfer function
dimensional analysis 1–2
  application to control problems 6–9
  scientific application 3–6
dimensional matrix 16–17
dimensional set 17–18
dimensional similarity 2–3, 5–6
  control: *see* control of dimensionally similar systems
  defining 19–20
  discrete time 46–8
  model law 21
  prototype and model systems 19–20
  scaling factors 20

state space 44–6
transfer function 41–4
dimensionless representation 3, 5
discrete time models 35–7
overview 23
properties 23
reduced effective transfer function (RETF) 67–8
state space 27–32
transfer function 23–7
transfer function *vs.* state space 32–5
dimensions
arithmetic: *see* arithmetic of dimensions
classification 11
defined 11
dependent and independent 14
derived 12
fundamental 11
power law monomial 14
system of units 13–14
discrete time control 111–13
discrete time dimensional similarity 46–8
sampled-data transfer function 46–7
state space 47–8
transfer function 46
discrete time models dimensionless representation 35–7
state space 36–7
transfer function 35–6
dynamical systems: *see* dimensional similarity; dimensionless representation

first order linear time-invariant plant, SMRAC for 126–9
Froude number 3, 23

homogeneity, of PID tuning rules 8–9
characterization 80–2
closed loop transfer functions 83–6
dimensionless controller parameters 80
dimensionless controller representation 83
nonhomogeneous rules 89–94
optimality 87–9

independent dimensions 14
indexes, comparative framework, PID tuning rules 104
integral 12
integral absolute error, comparative framework 104, 105
integral errors, comparative framework 100–2
international system (SI): *see* SI system
inverse response analysis, RETF 68–72
pole-zero cancelation 71
time delay factorization 70, 71
inverse response analysis, second order 52–64
see also second order inverse response model

mathematical equation 1
homogeneity and 2
matrix
dimensional 16–17
dimensional set 17–18
model reference adaptive control (MRAC) 121
monomial power law 14
MRAC: *see* model reference adaptive control (MRAC)

nonhomogeneous and homogeneous tuning rules 89–94
see also homogeneity, of PID tuning rules

optimality, of homogeneous tuning rules 87–9
weighting factors 88–9

partial similarity 113–15
physical equation 1
homogeneity and 2

physical quantity 11
physical relations 3–4
PID tuning rules
  dimensionless comparison 97–107
    see also comparative framework, PID tuning rules
  homogeneity of 79–94
    see also homogeneity, of PID tuning rules
pole-zero cancelation
  inverse response analysis, RETF 71
  reduced order model 73–4
polymerization reactor 76–7
power 12
power law monomial: *see* monomial power law
Prandtl number 3
product 12
proportional integral derivative (PID) tuning rules: *see* PID tuning rules

quotient 12

reduced effective transfer function (RETF)
  application examples 75–7
  dimensionless representation 67–8
  inverse response analysis 67–72
  problem statement 64–7
reduced order model 72–4
  pole-zero cancelation 73–4
  time delay factorization 73–4
Reynolds number 3

second order inverse response model 52–64
  application examples 61–4
  dimensionless representation 53–4
  identification procedure 54–61
  problem statement 52–3
sensitivity function bandwidth 103, 104
sensitivity peak function 104, 105
similar model reference adaptive control (SMRAC) 122, 123–6
  adaptive scheme 124–6
  for arbitrary order plants 129–33
  dimensionally similar models 123–4
  for first order linear time-invariant plant 126–9
single input/single output (SISO) 64
SISO: *see* single input/single output (SISO)
SI system 13
  derived units 13
  fundamental dimensions 13
SMRAC: *see* similar model reference adaptive control (SMRAC)
state space dimensional similarity 44–6
  discrete time 47–8
state space dimensionless representation 27–32
  discrete time models 36–7
  interpretation 30–2
  inverted pendulum 29–30
  linear time invariant (LTI) 27
  system transformation 28
  transfer function *vs.* 32–5
  transformations 30–2
system identification and model reduction 8
  application examples 61–4
  difficulties 51
  general procedure 51–2
  overview 51
  reduced effective transfer function reduction 64–77
  second order inverse response model identification 52–64
system of units 13–14

time delay factorization
  inverse response analysis, RETF 70, 71
  reduced order model 73–4
TITO: *see* two inputs/two outputs (TITO)

transfer function dimensional similarity 41–4
discrete time 46–7
sampled-data transfer function 46–7
transfer function dimensionless representation 23–7
dimensionless form 26–7
dimensionless numbers 25–6
discrete time models 35–6
parameters 24
vs. state space 32–5
two inputs/two outputs (TITO) 66

units 11
see also SI system

Vinante and Luyben distillation column 75–6